Introduction to Multisim®

for the DC/AC Course

Gary D. Snyder

Prentice Hall

Boston   Columbus   Indianapolis   New York   San Francisco   Upper Saddle River

Amsterdam   Cape Town   Dubai   London   Madrid   Milan   Munich   Paris   Montreal   Toronto

Delhi   Mexico City   Sao Paulo   Sydney   Hong Kong   Seoul   Singapore   Taipei   Tokyo

Editor in Chief: Vernon Anthony
Acquisitions Editor: Wyatt Morris
Editorial Assistant: Chris Reed
Director of Marketing: David Gesell
Marketing Assistant: Les Roberts
Senior Managing Editor: JoEllen Gohr
Project Manager: Rex Davidson
Senior Operations Supervisor: Pat Tonneman

Operations Specialist: Laura Weaver
Art Director: Candace Rowley
Cover Designer: Rachel Hirschi
Cover Art: Shutterstock
Lead Media Project Manager: Karen Bretz
Printer/Binder: Bradford & Bigelow, Inc.
Cover Printer: Lehigh-Phoenix Color
Text Font: Times Roman

LabVIEW, Multisim, NI, Ultiboard, and National Instruments are trademarks and trade names of National Instruments. Other product and company names are trademarks or trade names of their respective companies.

10 9 8 7 6 5 4 3

Prentice Hall
is an imprint of

www.pearsonhighered.com

ISBN 10: 0-13-508041-X
ISBN 13: 978-0-13-508041-2

Contents

Preface v

1. Introduction to Multisim 1

2. Schematic Capture with Multisim 9

3. Basic Electronics Concepts 23

4. DC Power Sources 33

5. Introduction to Resistive Circuits 43

6. Ohm's Law 53

7. Power in DC Circuits 57

8. Series Resistive Circuits 63

9. Kirchhoff's Voltage Law 69

10. Voltage Dividers 73

11. Parallel Resistive Circuits 77

12. Kirchhoff's Current Law 85

13. Loading Effects 89

14. Working with Circuit Equivalents 93

15. Series-Parallel Resistive Circuits 97

16. Superposition Theorem 101

17. Thevenin's Theorem 105

18. Norton's Theorem 111

19. Maximum Power Transfer 117

20. Branch Current Analysis 123

21. Loop Current Analysis 129

22. Node Voltage Analysis 135

23. **Bridge Circuits** 141

24. **Delta and Wye Circuits** 149

25. **The Oscilloscope** 155

26. **DC Response of RC Circuits** 163

27. **DC Response of RL Circuits** 169

28. **Introduction to Reactive Circuits** 181

29. **Introduction to Complex Circuits** 191

30. **AC Response of RC Circuits** 197

31. **AC Response of RL Circuits** 211

32. **AC Response of RLC Circuits** 223

33. **Power in Complex Circuits** 229

34. **Series Resonance** 233

35. **Parallel Resonance** 241

36. **Transformer Circuits** 245

37. **Low-Pass Filter Circuits** 253

38. **High-Pass Filter Circuits** 261

39. **Band-Pass Filter Circuits** 269

40. **Band-Stop Filter Circuits** 275

Preface

About This Workbook

The intent of this workbook is to help you develop a working knowledge of the National Instruments Multisim® software for entering and analyzing circuit designs that illustrate basic concepts in dc and ac electronics. Each section will contain some background theory for the circuits that you will investigate, but only to help provide context for the specific topics that the section will cover. For best results, you should use this workbook to supplement, rather than replace, a textbook that discusses the subject material in depth.

Workbook Organization

This workbook consists of 40 sections, each of which deals with a specific topic or electronics circuit application. Your instructor may have you work each section separately, or choose to merge sections that cover related topics (such as series resistive circuits and voltage dividers) into a single study topic.

A typical section will consist of some or all of the following subsections.

Introduction

The section introduction reviews basic electronics theory and concepts that relate to the topic in question, helps place the theory in context for the Multisim exercises that will follow, and provides specific learning objectives for the section.

Pre-Lab

The pre-lab contains tasks to help you prepare for completing the design verification and application exercise. These tasks include calculating circuit values you will measure in the design verification phase and familiarizing yourself with specific aspects of the Multisim software.

Design Verification

The design verification typically consists of using the Multisim software to simulate one or more sample circuits that accompany this workbook and comparing the values you measure with Multisim with your calculations from the design pre-lab.

Application Exercise

The application exercises provide hands-on opportunities to reinforce Multisim and electronics learning concepts.

Troubleshooting Exercise

The troubleshooting exercise helps you learn to use circuit theory and circuit measurements to identify common circuit problems.

Section Summary

The section summary reviews the main concepts in the section and relates them to concepts you will encounter in other workbook sections, more advanced studies, or real-world electronics applications.

Using This Workbook

Using Windows

This workbook assumes that you are familiar with using Microsoft® Windows® (starting applications, opening files, etc.) and with standard Windows operations and terminology (double-clicking icons, minimizing windows, etc.). If you have any questions about Windows, refer to the on-line help or standard Windows documentation.

Multisim Software

This workbook also requires you to have a copy of the Multisim software. Although the workbook sections reference and discuss Multisim 10, the interface and features in Multisim 9 are similar enough to those in Multisim 10 for you to use Multisim 9. This workbook references the sample circuit by filename only, but all files are available in both Multisim 9 (.MS9) and Multisim 10 (.MS10) formats. If a section references file "Ex01-01" you would open the file Ex01-01.MS9 if you are using the Multisim 9 software, and the file Ex01-01.MS10 if you are using the Multisim 10 software.

Note that you can use Multisim 10 to open and work with Multisim 9 circuits, but that the opposite is not true. In addition, Multisim 10 cannot save files in Multisim 9 format, even those that were originally in Multisim 9 format.

Workbook Conventions for Components

Bold italic font identifies both quantities, such as resistor R_1 or dc voltage supply V_S, and their value, such as "R_1 = 100 ohms" or "V_S = 12 V". You should be able to identify the specific meaning of the reference from its context.

Acknowledgments

I would like to express my great appreciation to David Buchla and Thomas Floyd for the benefit of their considerable authoring talents and experience and for the great opportunity of having worked with them in the past. I also wish to thank Wyatt Morris, Christopher Reed, Rex Davidson, and the technical staff of Pearson Education for their invaluable assistance in helping me to prepare this book for release, as well as the staff and instructors at ITT for their patience and understanding with the inevitable typos and "gotchas" that pop up in the early drafts of a book.

1.   Introduction to Multisim

1.1   Introduction

The world of electronics is continually changing. To remain competitive, electronics companies must rapidly design, prototype, test, manufacture, and market their products. In the past two-sided boards and leaded components were the standard and companies could use perfboard, protoboards, and wire-wrap boards to evaluate new parts and to build and test circuit prototypes. In modern electronics multilayer boards and surface-mount components are the standard, so creating physical prototypes from scratch is both problematic and expensive. To help solve this problem, component manufacturers often provide evaluation kits so that customers can determine whether a new part will meet their needs and better understand the practical requirements for using the part in circuits. Another valuable resource is specialized software to quickly design, evaluate, and modify circuit designs.

The Multisim software is a program that acts as a virtual electronics laboratory. You can use the Multisim program not only to create electronic circuits on your computer, but also to simulate ("run") the circuits and use virtual laboratory instruments to make electronic measurements.

In this section, you will:

- Learn how to start and close the Multisim application.

- Familiarize yourself with the Multisim workspace.

- Identify the various Multisim toolbars and their components.

1.2   Starting the Multisim Program

You can use standard Windows methods to start the Multisim program. You will learn one more way in the next section.

1.2.1   Starting the Multisim Program from the Windows Desktop

If you added a Multisim shortcut to your desktop during installation, then you can use the desktop shortcut to start the Multisim program. To do so:

1.   Navigate to the desktop.

2.   Double-click the Multisim icon. Your icon may differ from that of Figure 1-1.

Figure 1-1: Multisim Icon

1.2.2   Starting the Multisim Program from the Start Menu

Note that the following instructions assume that you accepted the default locations for your version of the Multisim software.

If you are using the Multisim 9 software:

1.   Click the **Start** button in the lower left corner of the Windows screen.

2.   Click **Programs**.

3.   Click the "Electronics Workbench" folder.

4. Click the "Multisim 9" folder.

5. Click "Multisim 9".

If you are using Multisim 10 software:

1. Click the **Start** button in the lower left corner of the Windows screen.

2. Click **Programs**.

3. Click the "National Instruments" folder.

4. Click the "Circuit Design Suite 10.0" folder.

5. Click "Multisim".

1.2.3 Starting the Multisim Program Using the Run… Command

Note that the following instructions assume that you accepted the default locations for your version of the Multisim software.

If you are using Multisim 9 software:

1. Click the **Start** button in the lower left corner of the Windows screen.

2. Click **Run…**.

3. Click the **Browse…** button.

4. Select the C: drive in the **Look in:** drop-down list.

5. Double-click the "Program Files" folder.

6. Double-click the "Electronics Workbench" folder.

7. Double-click the "EWB9" folder.

8. Double-click the "Multisim.exe" file.

Alternatively, you can enter "C:\Program Files\Electronics Workbench\EWB9\Multisim.exe" in the **Open…** text box and click "**OK**".

If you are using Multisim 10 software:

1. Click the **Start** button in the lower left corner of the Windows screen.

2. Click **Run…**.

3. Click the **Browse…** button.

4. Select the C: drive in the **Look in:** drop-down list.

5. Double-click the "Program Files" folder.

6. Double-click the "National Instruments" folder.

7. Double-click the "Circuit Design Suite 10.0" folder.

8. Double-click the "Multisim.exe" file.

Alternatively, you can enter "C:\Program Files\National Instruments\Circuit Design Suite 10.0\Multisim.exe" in the **Open…** text box and click "**OK**".

You may wonder why you would ever use the **Run…** command to start Multisim or any other Windows application. The answer is that once you have used this method the Windows operating system will remember the path to the Multisim program. You can then select Multisim directly from the drop-down list in the **Run…** window, unless your computer's security settings delete your session history.

1.3   The Multisim Interface

Once the Multisim program starts you will see the screen (or one much like it) shown in Figure 1-2.

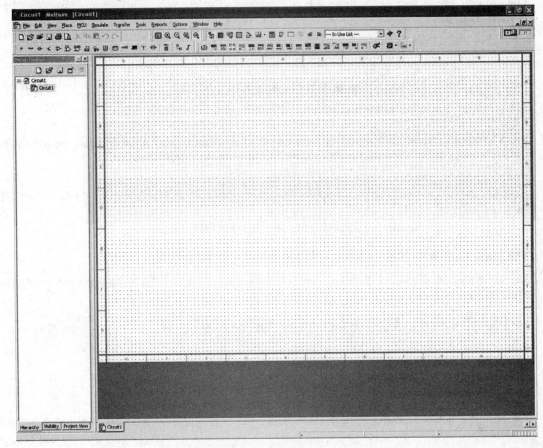

Figure 1-2: Multisim Interface

In addition to the standard Windows title bar, menu bar, toolbar, and screen controls, the Multisim interface contains a number of special toolbars that allow you to create and simulate circuits. The number and type of toolbars that you will see in the Multisim interface depends upon which toolbars you enable in the **View → Toolbars...** menu. The following sections provide a brief description of some of the toolbars that you will use in this workbook. The figures show toolbars from the Multisim 10 software.

1.3.1   The Title Bar

The **Title** bar is the region at the very top of the screen and is common to all Windows applications. The left side of the bar contains information about the application and file, while the right side of the bar contains controls to minimize, maximize, and close the application.

1.3.2   The Menu Bar

The **Menu** bar, shown in Figure 1-3, contains standard Windows menus (such as **File**, **Edit**, **View**, and **Help**) and menus that are specific to Multisim (such as **Place**, **MCU**, **Simulate**, and **Reports**). These menus let you configure and operate the application.

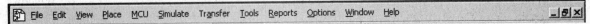

Figure 1-3: The Multisim Menu Bar

1.3.3 The Standard Toolbar

The **Standard** toolbar, shown in Figure 1-4, contains tools for performing common operations from the File and Edit menus, such as creating a new file, printing a file, and pasting information from the Clipboard into the open file.

Figure 1-4: The Standard Toolbar

1.3.4 The View Toolbar

The **View** toolbar, shown in Figure 1-5, contains tools for performing common Windows operations from the View menu, such as zooming in on or zooming out from the current view.

Figure 1-5: The View Toolbar

1.3.5 The Main Toolbar

The **Main** toolbar, shown in Figure 1-6, contains tools to access various Multisim features and information about the current circuit.

Figure 1-6: The Main Toolbar

1.3.6 The Simulation Run Toolbar

The **Simulation Run** toolbar, shown in Figure 1-7, contains tools that let you start, stop, pause, and resume a simulation.

Figure 1-7: The Simulation Run Toolbar

1.3.7 The Component Toolbar

The **Component** toolbar, shown in Figure 1-8, contains tools that let you access various components with which to create and analyze circuits.

Figure 1-8: The Component Toolbar

1.3.8 The Instruments Toolbar

The **Instruments** toolbar, shown in Figure 1-9, provides instruments with which you can measure and evaluate the operation of circuits. Some instruments, like the multimeter, oscilloscope, and logic analyzer, are real devices that technicians and engineers use to analyze real-world circuits. Other instruments, like the Bode plotter and logic converter, exist only within the Multisim application and are convenient tools for you to simulate, analyze, and debug your circuit designs.

Figure 1-9: The Instruments Toolbar

You may have noticed that the two tools on the far right of the instrument toolbar have small arrows pointing down next to them. These are "flyouts" that open menus that provide further selections for the tools.

1.3.9 Tool Tips

With so many toolbars and tools available the Multisim interface may seem rather confusing at first. To help you, the Multisim software uses tool tips to help you identify each tool. A tool tip is a small text box that tells you what the tool is. To activate a tool tip, simply let the pointer hover, or remain over, the tool you wish to identify. After a short time the tool tip for that tool will appear, as shown in Figure 1-10. The tool tip indicates that the highlighted instrument (the tool with the border around it) is the function generator.

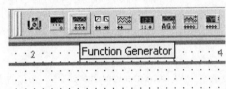

Figure 1-10: Function Generator Tool Tip

In the exercise in the next section you will use the tool tips to identify a number of tools and better familiarize yourself with some of the Multisim tools.

1.4 Section Exercise

1. Use one of the methods in Section 1.2 to start the Multisim program.

2. Use the information in Section 1.3 and the Multisim tool tips to complete Table 1-1 with

   - the toolbar on which the tool appears, and

   - the identity of the tool.

Table 1-1: Multisim Tool Identification

Tool Graphic	Associated Toolbar	Tool Identity

Tool Graphic	Associated Toolbar	Tool Identity

1.5   Closing the Multisim Program

There are two standard ways to close the Multisim program. If you made any changes to the circuit file (even if the net effect of those changes did not change the circuit file, such as adding a component and then deleting it), then the program will display a dialog box similar to that in Figure 1-11.

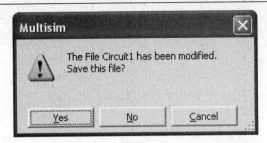

Figure 1-11: File Save Reminder Dialog Box

If you receive this notification, click the **No** button. You will learn about saving files in the next section.

1.5.1 Closing the Multisim Application from the File Menu

1. Click **File** in the Multisim menu bar.

2. Click **Close**.

1.5.2 Closing the Multisim Program with the Application Close Button

To use the **Application Close** button to close the Multisim application, click the **Close** (☒) button at the far right of the blue title bar. If you click the **Close** button in the Multisim menu bar, then you will close the circuit file but not the program.

1.6 Section Summary

The Multisim interface provides a variety of toolbars and tools for creating, modifying, and simulating circuit designs. Knowing which tools the Multisim software provides and where you can find them will greatly simplify working with circuit designs. In the following section you will learn how to use some of these tools to create, modify, and save your circuit designs.

2. Schematic Capture with Multisim

2.1 Introduction

A schematic is a graphical representation of a circuit design. Each symbol represents a specific type of component, and typically displays such electrical information as the component value, tolerance, and power rating. A component symbol can also show user-oriented information such as a reference designator and manufacturer part number, and manufacturing information that indicates the physical package or footprint.

One of the Multisim program's primary functions is to allow you to create schematics on your computer. This process is called **schematic capture** and closely resembles the process of building an actual circuit. Schematic capture consists of

- selecting the necessary components,
- arranging the components in the workspace, and
- connecting the components together to create the desired circuit.

Designs can range from simple circuits that consist of a few parts (such as a flashlight or transistor switch) to complicated circuits (like a multi-stage amplifier or digital state machine). Regardless of the size or complexity of the circuit, however, the basic schematic capture process is the same.

In this section you will:

- Construct a new circuit.
- Learn how to save a circuit file.
- Learn how to open an existing circuit file.
- Learn how to modify an existing circuit.

2.2 Constructing a New Multisim Circuit

In this section you will build a flashlight, or at least a schematic representation of one.

2.2.1 Placing Components

1. Use one of the methods from Section 1.2 to start the Multisim program.

2. Click the **Place Source** tool in the **Component** toolbar to open the component browser in Figure 2-1.

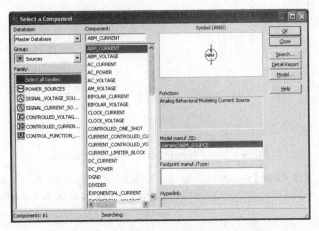

Figure 2-1: Sources Component Browser

3. Select POWER_SOURCES in the **Family** window and DC_POWER from the **Component** list. Note that the symbol in the window changes to that of a battery (i.e., a dc source).

4. Click the **OK** button. You will return to the circuit window in the Multisim interface. Note that as you move the pointer, the component symbol moves with it. Move the battery symbol to a position near the left center of the circuit window (refer to Figure 2-2) and click to place it.

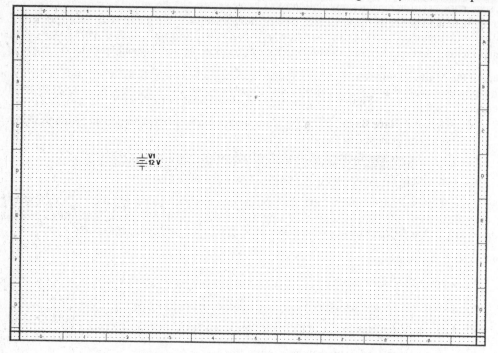

Figure 2-2: Circuit Window with Battery Placed

At this point, depending on your global preferences, the program may automatically return you to the component browser. If it does not, click the **Place Source** tool again.

5. In the component browser, select GROUND from the **Component** list.

6. Click the **OK** button and place the ground symbol beneath the battery symbol, as shown in Figure 2-3.

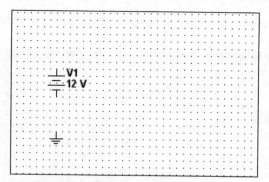

Figure 2-3: Circuit with Ground Placed

7. Place a second ground symbol to the right of the other symbols on the board, as shown in Figure 2-4. This completes the power source symbols you will need to construct the circuit.

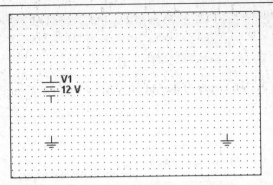

Figure 2-4: Circuit with Second Ground Placed

8. Click the **Place Basic** tool in the **Component** toolbar. The component browser appears with the basic components, as shown in Figure 2-5.

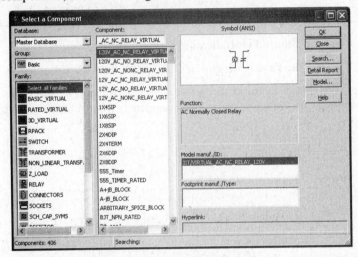

Figure 2-5: Basic Component Browser

9. Select SWITCH in the **Family** window and SPST from the **Component** list. If you do not see SPST in the component list, use the vertical scroll bar on the right side of the component list to scroll down until you see it.

10. Click the **OK** button and place the switch symbol above and to the right of the battery, as shown in Figure 2-6.

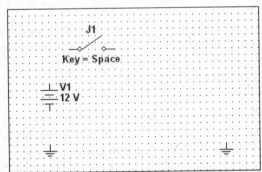

Figure 2-6: Circuit with SPST Switch Placed

11. Click the **Place Indicator** tool. The component browser appears with the indicators, as shown in Figure 2-7.

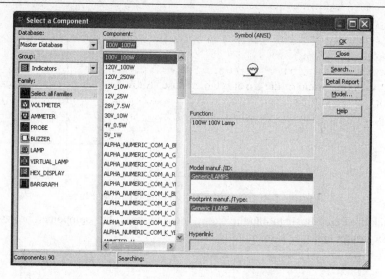

Figure 2-7: Indicators Component Browser

12. Select LAMP in the **Family** window and 12V_10W from the **Component** list.

13. Click the **OK** button and place the lamp to the right of the SPST switch, as shown in Figure 2-8. This completes the component placement segment of your schematic capture project.

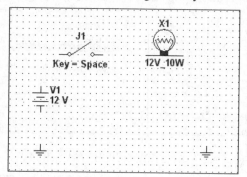

Figure 2-8: Circuit Window with Lamp Placed

2.2.2 Connecting Components

The Multisim program has no **Wire** tool to connect components. You create a wire by clicking on the start and end points for the wire.

Note that the program will not let you create a wire that does not begin on a component terminal or end on a wire segment or component terminal. If you wish to begin a wire segment on a wire segment or leave one end of a wire segment unconnected you must connect the wire to a special terminal called a junction. To place a junction, select **Junction** from the **Place** menu. When you end a wire on a wire segment the Multisim program automatically creates a junction for you.

1. Move the tip of the pointer near the cathode (negative terminal) at the bottom of the battery. The pointer will change from an arrow to a small crosshair, indicating that it is ready to make a connection to the battery terminal.

2. Click the left mouse button and drag the crosshair towards the left ground symbol. As you do so, you will see a wire being drawn. If you do not, position the pointer near the battery and left-click again when the crosshair is visible.

3. When you reach the terminal at the top of the ground symbol, left-click the mouse. This will connect the wire from the battery to the ground symbol, as shown in Figure 2-9. Depending on your program settings, you may or may not see the net name (in this case "0") for the wire. To show or hide the net name select **S̲heet Properties...** from the **O̲ptions** menu in the menu bar. In the **Net Names** field of the **Circuit** tab, choose "Show All" to show the names of all nets and "Hide All" to hide the names of all nets. "Use Net-specific Setting" will show only the nets that you individually configure to show the net name. This example used the "Show All" setting.

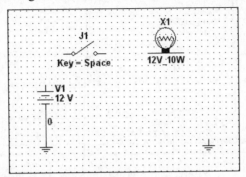

Figure 2-9: Circuit with Battery Connected to Ground

4. Finish connecting the remaining components, as shown in Figure 2-10.

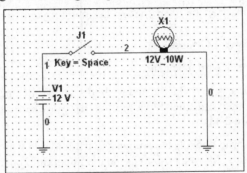

Figure 2-10: Circuit with Completed Connections

Congratulations! You've just completed your first Multisim circuit!

2.3 Saving a Multisim Circuit File

A good idea is always to save your work, especially after working for more than a few minutes. There are two methods for saving a Multisim circuit file. If you wish, use one of the standard methods below to save your circuit to your computer.

2.3.1 Using the Save File Tool to Save a Circuit File

1. Click the **Save File** tool on the standard toolbar.

2. If you have saved the file before, the Multisim program will use the same name and location as before to save the file. If you are saving the circuit for the first time the **Save As** window will appear, as shown in Figure 2-11.

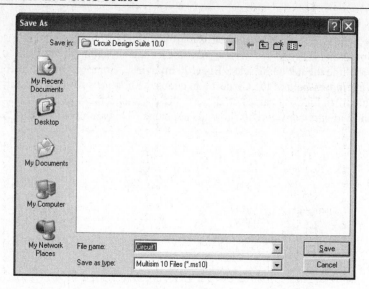

Figure 2-11: Multisim Save As Window

3. Use the **Save in:** drop-down list to navigate to the folder in which you wish to save the file, and enter the file name you wish to use in the **File name:** box.

4. Once you have finalized the file name and location for your circuit file, click the **Save** button.

2.3.2   Using the File Menu to Save a Circuit File

1. Click **File** on the standard menu bar.

2. If you have previously save the file and wish to save the file under the same name, click **Save** on the **File** menu. Otherwise, click **Save As...** on the **File** menu. The **Save As** window will appear, as shown in Figure 2-11.

3. Use the **Save in:** drop-down list to navigate to the folder in which you wish to save the file, and enter the name of the file you wish to use in the **File name:** box.

4. Once you have finalized the file name and location for your circuit file, click the **Save** button.

2.4   Opening an Existing Multisim Circuit

You can use one of the standard methods below to open a Multisim circuit that you or someone else created and saved. The Multisim 9 version can open circuit files created by Electronics Workbench 5 and Multisim versions 6 through 9. The Multisim 10 version can open circuit files created by Electronics Workbench 5 and Multisim versions 6 though 10.

2.4.1   Using the Open File Tool to Open a Circuit File

1. Click the **Open File** tool on the standard toolbar. The **Open File** window will appear, as shown in Figure 2-12.

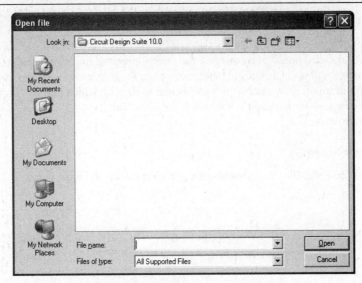

Figure 2-12: Multisim Open File Window

2. To specify the file you wish to open, you can

- use the **Look in:** and **Files of type:** drop-down lists to locate the file,

- use the **File name:** drop-down list to select the file from files you have previously opened, or

- enter the path and name of the file in the **File name:** box.

3. Once you have selected a file, click the **Open** button.

2.4.2 Using the File Menu to Open a Circuit File

1. Click **File** on the standard menu bar.

2. Click **Open...** on the **File** menu. The **Open File** window will appear, as shown in Figure 2-12.

3. To specify the file you wish to open, you can

- use the **Look in:** and **Files of type:** drop-down lists to locate the file,

- use the **File name:** drop-down list to select the file from files you have previously opened, or

- enter the path and name of the file in the **File name:** box.

4. Once you have selected a file, click the **Open** button.

2.4.3 Using Windows Explorer to Open a Circuit File

If the Multisim program is not running, you can use Windows Explorer to both start the Multisim program and open the circuit file.

1. Open Windows Explorer.

2. Navigate the folder that contains the circuit file you wish to open.

3. Double-click the circuit file. Windows will start the Multisim program, and the Multisim program will load the circuit file.

2.5   Modifying an Existing Circuit

Design is an ongoing process. At some point you will probably wish to change a circuit, whether to correct an error, move to the next phase of an incremental design process, improve a design, or just to tinker and see what the change will do. Changes to paper designs in the past were laborious and time-consuming. The Multisim program offers a number of features and tools with which you can quickly and easily change existing circuit designs. In this section you will learn some common ways to modify an existing circuit in the Multisim environment.

2.5.1   Open the Circuit to Modify

Open the circuit file Ex02-01. You should see the circuit shown in Figure 2-13.

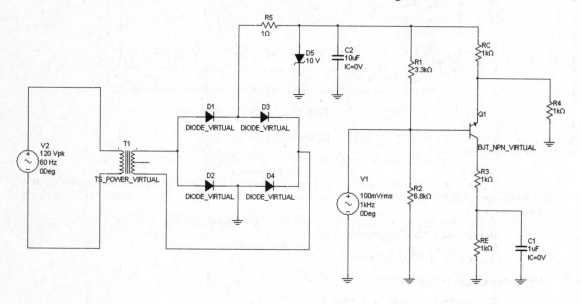

Section 2 MultiSIM Exercise Circuit 1

Figure 2-13: Practice Multisim Circuit

Don't worry if you don't understand the circuit—as shown it doesn't really do anything. For now you will concentrate on how to manipulate a circuit design rather than what the symbols represent or the intent of the circuit design.

Next, compare the circuit you opened with the reference circuit shown in Figure 2-14.

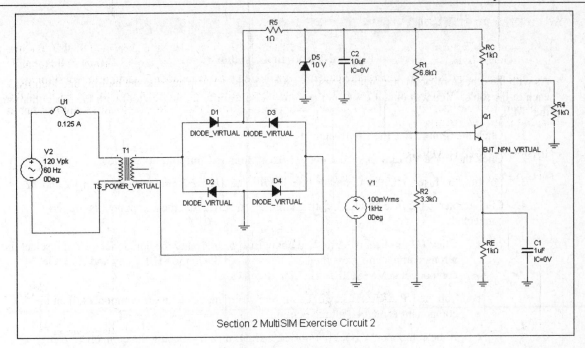

Section 2 MultiSIM Exercise Circuit 2

Figure 2-14: Reference Circuit

A careful comparison will reveal the following differences between the circuit file you opened and the reference circuit:

1. Addition of *U1* to the reference circuit

2. Orientation of *D2* and *D3*

3. Orientation of *D5*

4. Values of *R1* and *R2*

5. Orientation of *Q1*

6. Removal of *R3* from the reference circuit

7. Text at the bottom of the circuit

You will modify the circuit in Figure 2-13 to match that of the reference circuit. Before continuing, use **Save As**… to save the circuit file as Ex02-01_Mod.

If at any time you make a mistake, you can always use **Undo** in the **Edit** menu or use the **Undo** tool in the **Standard** toolbar to correct it.

2.5.2 Adding Components

Adding components is very similar to placing components. For this exercise you will add fuse *U1* and resistor *R3* from the **Virtual** toolbar. Unlike the other Multisim tools, the virtual tools are in blue-shaded boxes. If you cannot see the virtual tools in your Multisim interface, activate the **Virtual** toolbar as follows:

1. Click **View** on the standard menu bar.

2. Position your cursor over the **Toolbars** flyout.

3. When the toolbar menu opens, select **Virtual**. If **Virtual** is already checked, the **Virtual** toolbar should already be active and visible somewhere in the Multisim interface.

You should now see the **Virtual** toolbar, shown in Figure 2-15, in your Multisim interface.

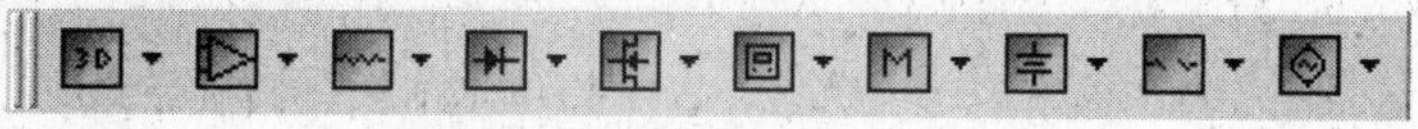

Figure 2-15: Virtual Toolbar

As with the other toolbars, you can hover your cursor over the tools to activate the tooltips and identify each of the tools. You will obtain *U1* from the **Show Misc Family** flyout (the tool with the box containing the "M").

1. Click the **Show Misc Family** flyout.

2. Click the **Place Virtual Fuse** tool in the **Miscellaneous Components** window.

3. Position the fuse so that its terminals line up with the top horizontal wire between *V2* and *T1*.

4. Click to merge the fuse into the existing circuit. To verify that the component is properly connected:

   a. Click *U1* to select it. A dashed box will appear around the component. If you select the wrong component, press the ESC key or hold down the SHIFT key and click the component again (SHIFT+click) to deselect it.

   b. Tap the UP ARROW key once. If the wiring moves with the component, then the component is merged into the circuit.

   c. If the component moves but not the wiring, then try placing the component again.

This completes the insertion of additional components into the circuit. Save your work before continuing.

2.5.3 Re-Orienting Components

Next, you will re-orient *D2*. To do so, you will first disconnect *D2* from the circuit and use the Multisim component right-click menu to "flip" the component so that it points in the opposite direction.

To disconnect *D2* from the circuit:

1. Click the wire segment to the left of *D2* to select it. The program will display solid boxes at the vertices of the wire to show that it is selected.

2. Next, select the wire segment to the right of *D2*. To do so, press and hold the SHIFT key and click the wire segment. The subcircuit should look like that of Figure 2-16.

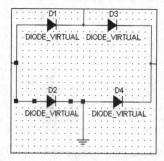

Figure 2-16: D2 Wire Segments Selected

3. Press the DELETE key to delete the wire segments. *D2* should now be disconnected from the rest of the circuit, as shown in Figure 2-17.

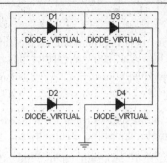

Figure 2-17: D2 Disconnected from Circuit

4. Right-click **D2**. A dashed box will appear around **D2** to show that it is selected, and the right-click menu shown in Figure 2-18 will appear.

Figure 2-18: Right-Click Menu

5. Click **Flip Horizontal** in the right-click menu. The right-click menu will close and the **D2** will be pointing in the opposite direction.

6. Place wires to reconnect **D2** back into the circuit.

Repeat the above procedure for **D3**, **D5**, and **Q1** to disconnect the component from the circuit, use the right-click menu to re-orient it, and route wires to reconnect it into the circuit. Note that you will use **Flip Vertical** in the right-click menu to re-orient **Q1**.

After re-orienting the necessary components, save your work before continuing.

2.5.4   Deleting Components

Next, you will delete **R3** from the circuit. Deleting a component is similar to deleting a wire segment. When you delete a component you will also delete wire segments that connect directly to the component, so you may need to re-route some wire segments.

1. Click **R3** to select it. The dashed selection block will appear around the component.

2. Press the DELETE key to delete **R3**.

3. Route a wire from **Q1** to the wire between **RE** and **C1** as shown in the reference circuit.

After deleting **R3**, save your work before continuing.

2.5.5   Changing Component Values

You could correct *R1* and *R2* by first deleting them and then placing the correct components into the circuit. With virtual components, however, you can correct the circuit simply by editing the component properties to change the component values.

To edit *R1*:

1.   Double-click *R1* to open the **Resistor** properties window. Alternatively, right-click *R1* to select it and open the right-click menu, and select **Properties…**.

2.   If you are using the Multisim 10 program, change the value in the **Resistance (R)** box from "3.3k" to "6.8k". If you are using the Multisim 9 program, change the value in the **Resistance:** box from "3.3" to "6.8" and leave the units as "kOhm" in the list box.

3.   Click the **OK** button.

Repeat the above procedure to change the value of *R2* from 6.8 kΩ to 3.3 kΩ. Save your work before continuing.

2.5.6   Editing Text

Editing text in Multisim is similar to editing text in other Windows applications. To edit the circuit description at the bottom of the schematic:

1.   Double-click the text. An insertion bar will appear at the start of the text.

2.   Use the RIGHT ARROW key repeatedly, or press the CTRL and END keys simultaneously, so that you can see the end of the text string.

3.   Use the cursor to highlight the number "1" at the end of the string.

4.   Type "2" to replace the highlighted text.

5.   Click away from the text to deselect it.

This completes the circuit modification exercise. Save your circuit file.

2.6   Application Exercise

Open the practice circuit Pr02-01. Modify the circuit so that it matches the circuit shown in Figure 2-19.

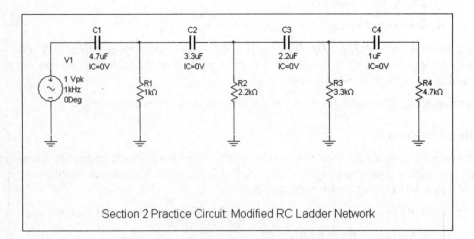

Figure 2-19: Modified Practice Circuit

2.7 Section Summary

This section has presented some of the tools and techniques with which you can create, modify, and save circuit designs. Up to this point, however, you have used the Multisim program as little more than a specialized drawing program, and the circuits have been little more than static drawings. In the next section you will learn to work with components and circuits as dynamic entities that model the behavior of their real-world counterparts.

3. Basic Electronics Concepts

3.1 Introduction

You can describe the operation of all electronic circuits in terms of basic electrical quantities. Some electrical quantities, such as voltage, are probably familiar to you. Other quantities, such as reactance, are probably not. Electronic components provide a practical means for you to work with electrical quantities so that you can design and construct a circuit. A battery, for example, can supply the voltage to a circuit, and resistors can determine how much current will flow in specific parts of the circuit. Once you understand electrical quantities you can understand the role of specific components and the operation of circuits containing them.

In addition to understanding what electrical quantities are you must also have some way to measure them. Designing a 12-volt power supply is pointless (and probably impossible) if you cannot determine how much a "volt" is. You must be able to measure electrical quantities like voltage and current to also determine whether or not a circuit is functioning correctly, and isolate any problems in the circuit.

In this section you will:

- Review basic electrical quantities and the units for each.

- Learn the components that are associated with electrical quantities.

- Learn measurement techniques for voltage, current, and resistance.

3.2 Electrical Quantities and Units

3.2.1 Charge

Charge, typically represented by Q, is measured in **coulombs** (C) and is an inherent property of matter, which consists of protons, electrons, and neutrons. Each proton carries a positive electrical charge, and each electron carries a negative electrical charge. Neutrons carry no net electrical charge. The charges on protons and electrons are equal in magnitude but opposite in sign so that an atom with an equal number of protons and electrons has no net electrical charge.

Like electrical charges will repel and opposite electrical charges will attract. A particle with a negative charge (such as an electron) will move towards a positive electrical charge and move away from a negative electrical charge, just as a particle with a positive charge (such as a positive ion) will move towards a negative electrical charge and move away from a positive electrical charge. Because of this phenomenon, you must apply energy in the form of work to a system to separate positive and negative charges. Conversely, opposite charges moving together will perform work and return energy to the system. You can liken this to two objects connected by an elastic band. You must perform work to pull the objects apart, and the energy you expend is stored in the elastic band. When you release the objects, the energy in the elastic band is converted to kinetic energy as the objects move closer together again.

Although you will not often measure or work directly with electrical charge, it is fundamental to other electrical quantities.

3.2.2 Voltage

Voltage, typically represented by V and more rarely by E, is measured in **volts** (V). Separating particles with opposite charges requires work, and this separation increases the potential energy of the charges. Voltage represents the potential energy of the separated charges and is equal to their difference in potential energy per charge. A common analogy for voltage is that of water pressure in a gravity-fed water system. Energy in the form of work raises the water in the system to a storage tank. The water that is stored in the tank exerts pressure and possesses potential energy that can perform work when released.

Voltage sources apply a specific amount of voltage to a circuit. There are two basic types of voltage sources. One is the ac voltage source, in which the polarity of the voltage periodically reverses. The other is the dc voltage source, in which the polarity of the voltage does not change. Figure 3-1 shows the symbols for ac and dc voltage sources.

V1
120 Vrms
60 Hz
0°

V2
12 V

AC VOLTAGE SOURCE   DC VOLTAGE SOURCE

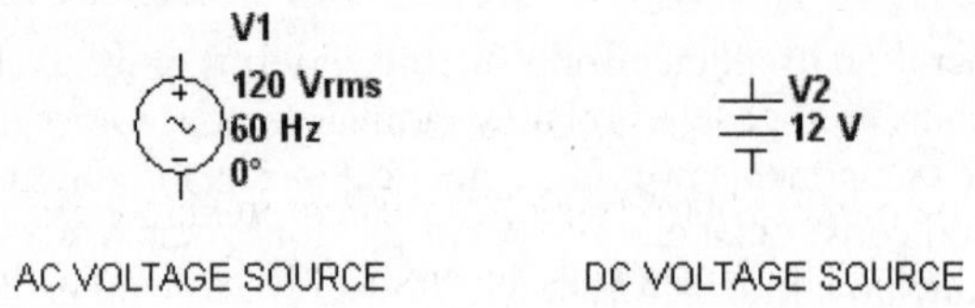

Figure 3-1: AC and DC Voltage Source Symbols

3.2.3 Current

Current, typically represented by *I*, is measured in **amperes** (A), commonly shortened to "amps." Electrical current is the net movement of charge. Just as you can think of voltage as the electrical pressure in a circuit, current is the electrical flow that results.

Although voltage sources typically generate currents you will sometimes encounter current sources in circuit designs. Current sources are similar to voltage sources, but inject a specific amount of current into a circuit rather than apply a specific amount of voltage to it. Figure 3-2 shows the symbols for ac and dc current sources.

I1
1 A
1kHz
0°

I2
1 A

AC CURRENT SOURCE   DC CURRENT SOURCE

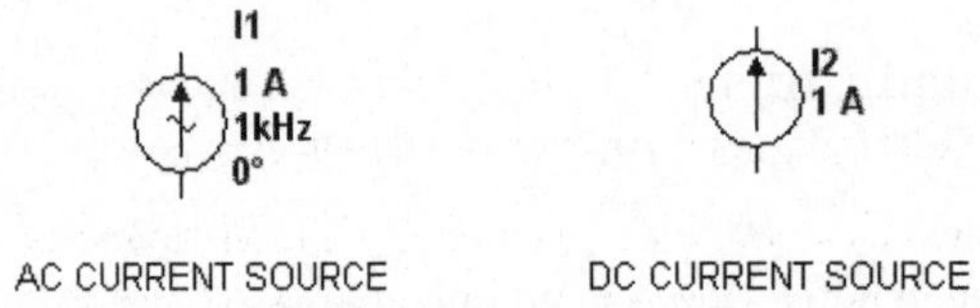

Figure 3-2: AC and DC Current Source Symbols

3.2.4 Resistance

Resistance, typically represented by *R*, is measured in **ohms** (Ω). Resistance is the opposition to current. Just as the diameter of a pipe determines how much water will flow for a given amount of water pressure, resistance determines how much current will flow for a given voltage.

The **resistor** provides a known amount of resistance in a circuit. Resistors can be either fixed or variable, as shown in Figure 3-3. The symbol on the left shows a fixed resistor, which is a two-terminal device whose value cannot change. The symbol on the right shows a three-terminal device called a potentiometer, sometimes mistakenly referred to as a variable resistor. The resistance value between the top and bottom terminals is fixed. The resistance between the third terminal (called the wiper) and the top or bottom terminal depends upon its physical position, which the user can vary. As the wiper moves towards the top terminal, the resistance between the top terminal and wiper decreases and the resistance between the bottom terminal and wiper increases. Conversely, as the wiper moves towards the bottom terminal the resistance between the top terminal and wiper increases and the resistance between the bottom terminal and wiper decreases.

R1
1.0kΩ

R2
1.0kΩ 50%
Key=A

FIXED RESISTOR   POTENTIOMETER

Figure 3-3: Resistor Symbols

If you tie the wiper to one of the other terminals, as shown in Figure 3-4 , you will create a two-terminal variable resistor. The resistance between the top and bottom terminals will change as the position of the wiper changes.

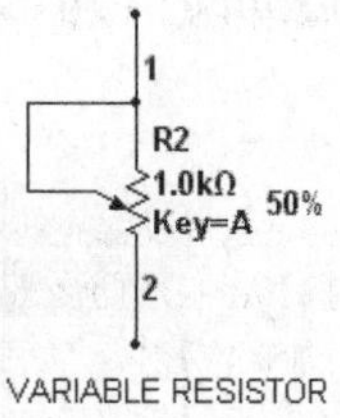

1

R2
1.0kΩ 50%
Key=A

2

VARIABLE RESISTOR

Figure 3-4: Potentiometer Connected as a Variable Resistor

3.2.5   Capacitance

Capacitance, typically represented by C, is measured in **farads** (F). Capacitance is the ratio of stored charge to the amount of voltage required to store the charge. A larger capacitance requires less voltage to store the same amount of charge than a smaller capacitance.

The **capacitor**, once known as a condenser, provides a known amount of capacitance in a circuit. As with resistors, capacitors come in both fixed and variable versions. Figure 3-5 shows the symbols for these.

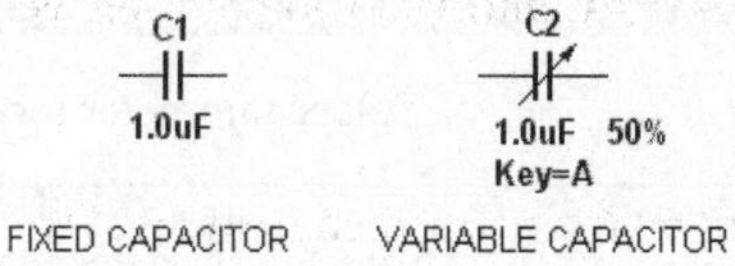

C1
1.0uF

C2
1.0uF 50%
Key=A

FIXED CAPACITOR   VARIABLE CAPACITOR

Figure 3-5: Fixed and Variable Capacitor Symbols

3.2.6   Inductance

Inductance, typically represented by L, is measured in **henries** (H). Inductance is the ratio of voltage to the rate of change in current required to induce the voltage. A larger inductance requires current to change less rapidly to produce the same voltage than a smaller inductance.

The **inductor**, sometimes called a coil or a choke, provides a known amount of inductance in a circuit. Figure 3-6 shows the symbols for fixed and variable inductors.

L1
1.0mH

L2
1.0mH 50%
Key=A

FIXED INDUCTOR   VARIABLE INDUCTOR

Figure 3-6: Fixed and Variable Inductor Symbols

3.2.7 Power

Power, typically represented by P, is measured in **watts** (W). Power is the rate at which energy is used, so that using the same amount of energy faster results in more power. In electrical circuits, power is equal to the product of current and voltage.

3.3 Multimeter Circuit Measurements

There are instruments to measure every electrical quantity, but the most common circuit measurement tool is the multimeter. Basic multimeters combine the functions of a voltmeter, ammeter, and ohmmeter to measure voltage, current, and resistance, respectively. More expensive multimeters can measure capacitance, test diodes, and perform other useful functions. The Multisim program provides a Multimeter tool on the **Instrument** toolbar, shown in Figure 3-7 in both minimized and expanded form. Table 3-1 summarizes the multimeter functions.

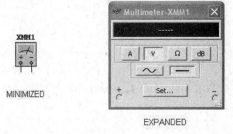

Figure 3-7: Multisim Multimeter Views

Table 3-1: Multimeter Functions

Button	Function
A	Selects ammeter mode.
V	Selects voltmeter mode.
Ω	Selects ohmmeter mode.
dB	Selects decibel mode.
∼	Selects ac measurement mode.
—	Selects dc measurement mode.
Set...	Opens multimeter settings dialog box.

3.3.1 Measuring Circuit Voltage

You always make voltage measurements with respect to some reference point. Usually this reference point is circuit ground, but need not be. A common convention is to specify a voltage at point X with respect to ground as V_X, and the voltage at point X with respect to point Y as V_{XY}. Voltage across a component is often represented as V_{REFDES}, where REFDES is the reference designator for the component. In Figure 3-8 XMM1 is measuring V_A, the voltage at point A with respect to ground, XMM3 is measuring V_B, which is the voltage at point B with respect to ground, and XMM2 is measuring V_{AB}, which is the voltage at point A with respect to the voltage at point B, or $V_A - V_B$. This voltage is also referred to as V_{R1}, as this is the voltage across **R1**.

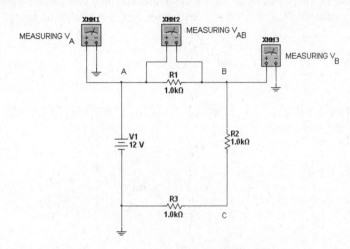

Figure 3-8: Voltage Measurement Conventions

The circuit must be powered when you make voltage measurements. Also, when making voltage measurements, you place the meter across components, and not between them. This means that voltage measurements are very convenient, as you do not have to disassemble a circuit to make them.

3.3.2 Measuring Circuit Current

The circuit must also be powered when you make current measurements. Unlike voltage measurements, current measurements require you to insert the multimeter into the circuit. This is because current must pass through the meter for the meter to measure it. NEVER connect an ammeter across a component to measure current, as very high current could flow through and damage the meter. Refer to Figure 3-9 for the correct and incorrect ways to connect an ammeter to a circuit.

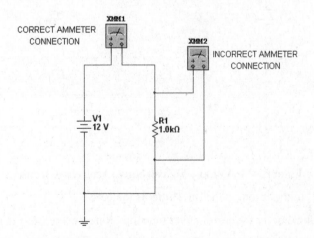

Figure 3-9: Correct and Incorrect Meter Connections for Current Measurements

A common convention is to identify the current through a component as I_{REFDES}, where REFDES is the reference designator for the component. In Figure 3-9 the current through **R1** would be I_{R1}.

3.3.3 Measuring Circuit Resistance

You must remove power from the circuit when making resistance measurements, and will typically disconnect the power supply from the circuit as well. You connect the meter across the points whose resistance you wish to measure. As with voltage measurements, a common convention is to identify the measured resistance as R_{XY}, where X and Y are the points to which you connect the meter. The multimeter in Figure 3-10 is measuring R_{DE}. As you will learn later, this is not the same as $R5$.

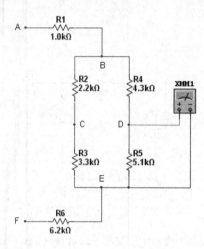

Figure 3-10: Measuring Resistance R_{DE}

3.4 Application Exercises

3.4.1 Measuring Circuit Voltages

1. Open the circuit Ex03-01 shown in Figure 3-11.

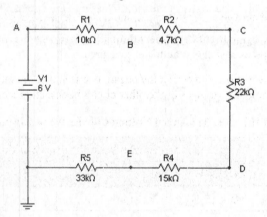

Figure 3-11: Voltage Measurement Exercise Circuit

2. Click the Multimeter tool in the **Instruments** toolbar.

3. Set the multimeter for the dc voltmeter function. Refer to Table 3-1 to review the multimeter functions.

4. Double-click the multimeter to expand the multimeter. Move the expanded view out of the way if it blocks your view of the circuit.

5. Click the negative (−) terminal of the multimeter and draw a wire to the ground node to connect the negative terminal to the ground node.

6. Click the positive (+) terminal of the multimeter and draw a wire to node A to connect the positive terminal to node A. If you wish, you can select the minimized multimeter tool and use the right-click menu to flip the multimeter and simplify the wiring, as shown in Figure 3-12.

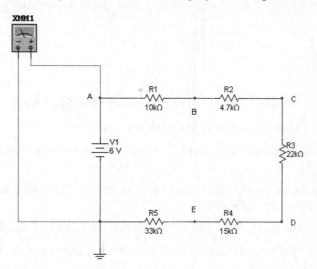

Figure 3-12: Connecting the Multimeter to the Voltage Measurement Exercise Circuit

7. Click the **Run** switch in the **Simulation Run** toolbar to activate the circuit. After a slight delay the expanded meter will show a voltage reading. The reading should be 6 V. This voltage is already recorded in the appropriate cell of Table 3-2.

8. Click the **Run** switch to deactivate the circuit.

9. Delete the wire between the multimeter and node A to disconnect the meter from the circuit.

10. Repeat Steps 6 through 9 for nodes B through E and record the voltages for each measurement to complete the top row of Table 3-2.

11. Complete Table 3-2 by connecting the negative terminal of the multimeter to nodes A through E in turn, measuring the voltages at the other circuit nodes, and recording the values.

Table 3-2: Measured Voltages for Exercise Circuit

		Measurement Node (Positive Multimeter Terminal)					
		GROUND	A	B	C	D	E
Reference Node (Negative Multimeter Terminal)	GROUND	0 V	6 V				
	A		0 V				
	B			0 V			
	C				0 V		
	D					0 V	
	E						0 V

3.4.2 Measuring Circuit Currents

1. Open the circuit Ex03-02 shown in Figure 3-13.

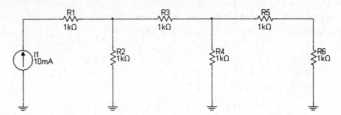

Figure 3-13: Current Measurement Exercise Circuit

2. Click the Multimeter tool in the **Instruments** toolbar.

3. Set the multimeter for the dc ammeter function. Refer to Table 3-1 to review the multimeter functions.

4. Double-click the multimeter to expand the multimeter. Move the expanded view out of the way if it blocks your view of the circuit.

5. Delete the wire between the top of the dc current source and resistor **R1**.

6. Click the negative (−) terminal of the multimeter and draw a wire to the left side of **R1**.

7. Click the positive (+) terminal of the multimeter and draw a wire to node A to connect the positive terminal to node A, as shown in Figure 3-14.

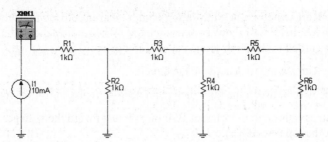

Figure 3-14: Connecting the Multimeter to the Current Measurement Exercise Circuit

8. Click the **Run** switch in the **Simulation Run** toolbar to activate the circuit. After a slight delay the expanded meter will show a voltage reading. The reading should be about 10 mA. This voltage is already recorded in the appropriate cell of Table 3-3.

9. Click the **Run** switch to deactivate the circuit.

10. Delete the wires between the multimeter and the circuit to disconnect the meter from the circuit.

11. Replace the wire you deleted to connect the multimeter to the circuit.

12. Repeat Steps 5 through 11 for the currents through **R2** through **R6**.

Table 3-3: Measured Currents for Exercise Circuit

Current	I_{R1}	I_{R2}	I_{R3}	I_{R4}	I_{R5}	I_{R6}
Measurement	9.997 mA					

3.4.3 Measuring Circuit Resistances

1. Open the circuit Ex03-03 shown in Figure 3-15.

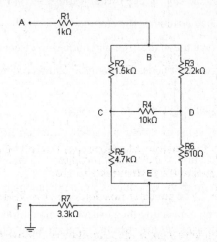

Figure 3-15: Resistance Measurement Exercise Circuit

2. Click the Multimeter tool in the **Instruments** toolbar.

3. Set the multimeter for the ohmmeter function. Note that the Multisim program automatically changes the voltage mode to the dc mode when you select the ohmmeter function. Refer to Table 3-1 to review the multimeter functions.

4. Double-click the multimeter to expand the multimeter. Move the expanded view out of the way if it blocks your view of the circuit.

5. Click the positive (+) terminal of the multimeter and draw a wire to node A to connect the positive terminal to node A.

6. Click the negative (−) terminal of the multimeter and draw a wire to node B to connect the negative terminal to node B, as shown in Figure 3-16.

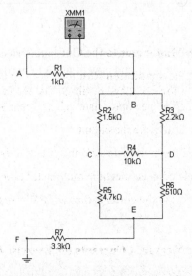

Figure 3-16: Connecting the Multimeter to the Resistance Measurement Exercise Circuit

7. Click the **Run** switch in the **Simulation Run** toolbar to activate the circuit. After a slight delay the expanded meter will show a resistance reading. The reading should be 1 kΩ. This resistance is already recorded in the appropriate cell of Table 3-4.

8. Click the **Run** switch to deactivate the circuit.

9. Delete the wire between the multimeter and node B to disconnect the meter from the circuit.

10. Repeat Steps 6 through 9 for nodes C through F and record the resistances for each measurement to complete the top row of Table 3-4.

11. Redraw the wire for the left side of **R1**. To do so:

   a. Click **Place** on the menu bar and select **Junction** from the drop-down menu.

   b. Move the junction to the right of the letter "A" and click.

   c. Draw a wire from the left side of **R1** to the junction.

12. Complete Table 3-4 by connecting the positive terminal of the multimeter to nodes B through F in turn, measuring the resistances between the other circuit nodes, and recording the values.

Table 3-4: Measured Resistances for Exercise Circuit

		Measurement Node (Positive Multimeter Terminal)					
		A	B	C	D	E	F
Reference Node (Negative Multimeter Terminal)	A	0 Ω	1 kΩ				
	B		0 Ω				
	C			0 Ω			
	D				0 Ω		
	E					0 Ω	
	F						0 Ω

3.5 Section Summary

This section introduced basic electronics concepts and common electronic components, including the resistor. It also introduced the multimeter and how to use it to measure voltage, current, and resistance in basic resistive circuits. You will use the multimeter extensively as you progress in electronics, both in future sections in this workbook and in the real world.

4.   DC Power Sources

4.1   Introduction

All circuits, from the simple flashlight you created in Section 2.2 to the most complex computers, require a source of power to operate. Virtually all circuits use a voltage source. An ideal voltage source will maintain a fixed voltage across its terminals regardless of the current it must provide (source) to a load. The polarity of a dc voltage source does not change, so that one terminal is always positive with respect to the other terminal. Common dc power sources include batteries and electronic lab power supplies.

Current sources can also provide power to a circuit. An ideal current source will maintain a fixed current regardless of the resistance of the circuit to which it connects. The polarity of a dc current source does not change, so that one terminal always sources current and the other terminal always sinks current. Current sources can be subcircuits that power other portions of a larger circuit or commercial constant current power supplies.

In this section you will:

*   Learn how to work with Multisim dc voltage and current sources.

*   Learn how to modify properties of circuit components.

*   Learn the characteristics of dc voltage and current sources.

4.2   DC Voltage Sources

4.2.1   Voltage of an Ideal Voltage Source

Open the circuit file Ex04-01 shown in Figure 4-1.   Before beginning the demonstration, you will use the component properties windows to make some changes. First you will modify the voltage source.

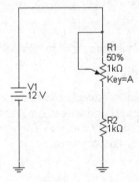

Figure 4-1: Voltage Source Demonstration Circuit

1.   Save the file as Ex04-01_Mod.

2.   Right-click the voltage source and select **Properties** from the right-click menu. The **DC_POWER** properties window will open.

3.   Select the **Value** tab, if it is not already selected.

4.   Change the value in the **Voltage (V)** box to 10. Leave the units in the list box as "V" (volts).

5.   Select the **Label** tab.

6.   Change the value in the **RefDes** box from the default of "V1" to "VS". Enter "Ideal Source" in the **Label** box.

7. Click the **OK** button.

8. Save your file to keep your changes up to this point. Your circuit should now look like Figure 4-2.

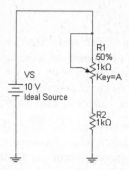

Figure 4-2: Demonstration Circuit with Modified Voltage Source

As you can see, changing the value in the **Voltage** box of the **Value** tab has changed the voltage of the dc voltage source. Similarly, the reference designator is now *VS* (a common designation for a voltage source) and the voltage source has a label that describes it. Next, you will change some properties of potentiometer *R1*.

1. Right-click the potentiometer and select **Properties** from the right-click menu. The **Potentiometer** properties window will open.

2. Select the **Value** tab, if it is not already selected.

3. If you are using the Multisim 10 software, click the down arrow to the right of the **Resistance (R)** box and select "10k" from the drop-down list. You can also select the value "1k" in the box and type "10k" to replace it. If you are using the Multisim 9 software, change the **Resistance:** value to "10" and leave the units as "kOhm" in the list box.

4. Change the value in the **Increment** box from "5" to "10".

5. Select the **Label** tab.

6. Change the reference designator from "R1" to "RVAR".

7. Add the label "Variable Load".

8. Click the **OK** button.

9. Save your circuit to keep your changes up to this point. Your circuit should now look like Figure 4-3.

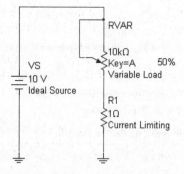

Figure 4-3: Demonstration Circuit with Modified Potentiometer

The modifications you made to the potentiometer had two visible effects, namely changing the reference designator to "RVAR" and adding the label "Variable Load". You may already have guessed what changing the increment value from 5% to 10% will do, but if not it will become clear when you simulate the circuit. Now you will change some properties of the fixed resistor.

1. Open the **Resistor** properties window for the fixed resistor.

2. Select the **Value** tab is it is not already selected.

3. If you are using the Multisim 10 software, change the value in the **Resistance (R)** box from "1k" to "1". If you are using the Multisim 9 software, leave the value in the **Resistance:** box as "1" and select "Ohm" from the list box.

4. Select the **Label** tab.

5. Change the reference designator from "R2" to "R1".

6. Add the label "Current Limiting".

7. Click the **OK** button.

8. Save your circuit to keep your changes up to this point. Your circuit should now look like Figure 4-4.

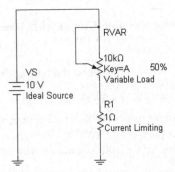

Figure 4-4: Demonstration Circuit with Modified Fixed Resistor

The changes to the circuit should not surprise you, although you might wonder why you changed the reference designator from "R2" to "R1". The only reason is that an "R2" reference designator in a circuit implies the existence of an "R1" reference designator, but R1 ceased to exist after you changed the reference designator of the potentiometer to "RVAR". Note that you could not first change the reference designator of the fixed resistor and then that of the potentiometer, as the Multisim program would not permit both components to have an "R1" reference designator at the same time, even temporarily.

If you are using Multisim 10 you may have noted that when your cursor passed over *RVAR* a slider appeared below the potentiometer. Place your cursor over *RVAR* now. When the slider appears, place the cursor on the slider and move to the right. As you move the slider the 50% to the right of the potentiometer will increase in 10% increments. Had you left the value in the **Increment** box at "5" this value would have increased in 5% increments. When the slider is fully to the right the value will be 100%. This means that the slider is at the top of the potentiometer and the potentiometer is placing 100% × 10 kΩ = 10 kΩ of resistance in the circuit.

Finally, add some devices to measure the dc supply voltage and circuit current. Although you could use the multimeter, the Multisim program offers some convenient monitoring devices that will measure voltage and current.

1. Click the **Show Measurement Family** flyout in the **Virtual** toolbar. The **Measurement Components** window will appear.

2. Click the **Place Ammeter (horizontal)** tool and place the ammeter in the top horizontal wire segment.

3. Click the **Place Voltmeter (vertical)** tool and place the voltmeter across the dc voltage source.

4. Save your circuit to keep your changes up to this point. Your circuit should look like that shown in Figure 4-5.

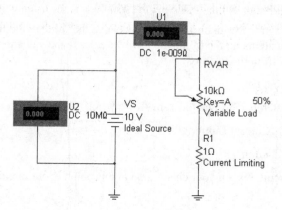

Figure 4-5: Demonstration Circuit with Measuring Devices

You are now ready to simulate the circuit.

1. If the **RVAR** percentage is not 100%, press the "A" key repeatedly until the percentage is 100%.

2. Click the **Run** switch to start the simulation.

3. Note the voltage and current values for the potentiometer setting (this may take a few seconds). The reading for the dc voltage source will be 10.000 V and the current reading will be 0.997 mA. These values are recorded for the 100% potentiometer setting in Table 4-1.

4. Reduce the potentiometer value by 10% by pressing SHIFT + A (holding down the SHIFT key and pressing the "A" key). The legend "Key=A" means that pressing "A" will increase the potentiometer value by one increment and pressing SHIFT + A will decrease the potentiometer value by one increment.

5. Record the voltage and current values for this potentiometer setting in Table 4-1.

6. Repeat Steps 3 and 4 for the remaining potentiometer settings.

7. Click the **Run** switch to stop the simulation.

Table 4-1: Circuit Measurements

Potentiometer Setting (%)	DC Source Voltage (V)	Circuit Current (A)	Potentiometer Setting (%)	DC Source Voltage (V)	Circuit Current (A)
100	10.000	0.998 m	40		
90			30		
80			20		
70			10		
60			0		
50					

Note that the dc source voltage remains constant as the circuit draws more current. This is what you should expect for an ideal voltage source.

4.2.2 Multiple Sources in a Circuit

Circuits can contain more than one power source, provided that the circuit does not violate certain rules. For voltage sources, one rule is that you cannot connect one terminal of the source directly to the other. This is not hard to understand, as this would put the terminals at the same voltage, which violates the nature of a voltage source. Another rule is that you cannot connect one voltage source in parallel with (directly across) another voltage source. Refer to Figure 4-6.

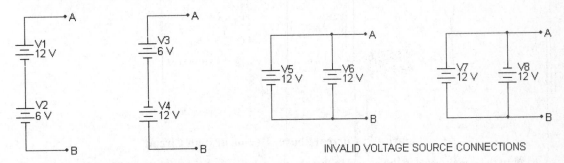

VALID VOLTAGE SOURCE CONNECTIONS

Figure 4-6: Multiple Voltage Source Connections

It is not hard to see why the two invalid voltage source connections are invalid. In both cases the 12V supply will try to keep the voltage between points A and B at 12 V, while the 6V supply will try to keep the voltage at 6 V. It is clearly impossible to determine the voltage between points A and B for ideal voltage sources, and for practical voltage sources these connections will result in excessive current flow. NEVER directly connect voltage sources in parallel. If you attempt to do so in a Multisim circuit then the software will warn you of the problem.

For the valid voltage source connections, called series connections, the voltage across A and B will be the algebraic sum of the voltage sources. To understand this, consider what happens as you start from B and work towards A. For the first circuit, $V2$ will keep the voltage at the junction of $V1$ and $V2$ 6 V more positive than V_B. Similarly, $V1$ will keep V_A 12 V more positive than the voltage at the junction of $V1$ and $V2$. The result is that V_A is 6 V + 12 V = 18 V more positive than V_B so that $V_{AB} = +18$ V. For the second circuit $V4$ will keep the voltage at the junction of $V3$ and $V4$ 12 V more negative than V_B, and V3 will keep V_A 6 V more positive than the voltage at the junction at $V3$ and $V4$. Consequently V_A is $(-12$ V$) + 6$ V $= -6$ V relative to V_B, so that $V_{AB} = -6$ V.

To verify this, open the circuit file Ex04-02, measure the voltage between points A and B, and record the results in Table 4-2. Remember when measuring the voltages to connect the + terminal to point A and the − terminal to point B.

Table 4-2: Measured V_{AB} Values

	Circuit 1 (V)	Circuit 2 (V)
Calculated	+18	−6
Measured		

4.3   DC Current Sources

4.3.1   Current of an Ideal Current Source

Open the circuit file Ex04-03 shown in Figure 4-7.

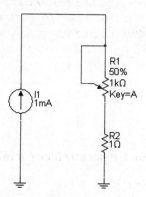

Figure 4-7: Current Source Demonstration Circuit

Modify the components in the circuit so that they match those of Figure 4-8. Although you can open the component properties window by selecting the necessary components and right-clicking, as you did in Section 4.2.1, the Multisim program offers a shortcut. You can open the components window by double-clicking on the components you wish to edit. Use this shortcut now to edit the properties of *I1*, *R1*, and *R2* so that your circuit matches that of Figure 4-8. Set the value in the **Increment** box for the potentiometer to 10.

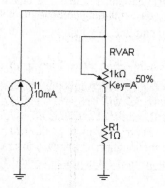

Figure 4-8: Modified Current Source Demonstration Circuit

Place a voltmeter across the current source *I1* and an ammeter between the current source and *RVAR* as shown in Figure 4-9.

Section 4 - DC Power Sources

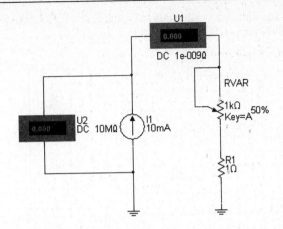

Figure 4-9: Current Source Demonstration Circuit with Measuring Devices

You are now ready to simulate the circuit.

1. If the **RVAR** percentage is not 100%, press the "A" key repeatedly until the percentage is 100%.

2. Click the **Run** switch to start the simulation.

3. Note the voltage and current values for the potentiometer setting (this may take a few seconds). The reading for the dc current source will be 10.008 V and the current reading will be 9.999 mA. These values are recorded for the 100% potentiometer setting in Table 4-3.

4. Press SHIFT + A to reduce the potentiometer value by 10%.

5. Record the voltage and current values for this potentiometer setting in Table 4-3.

6. Repeat Steps 3 and 4 for the remaining potentiometer settings.

7. Click the **Run** switch to stop the simulation.

Table 4-3: Circuit Measurements

Potentiometer Setting (%)	DC Source Voltage (V)	Circuit Current (mA)	Potentiometer Setting (%)	DC Source Voltage (V)	Circuit Current (mA)
100	10.008	9.999	40		
90			30		
80			20		
70			10		
60			0		
50					

Note that the dc source current remains constant but that the current source voltage decreases as the circuit resistance decreases. This is what you should expect for an ideal current source, as the current source requires less "pressure" to push 10 mA of current through the circuit as the resistance decreases.

4.3.2 Multiple Current Sources in a Circuit

As with voltage sources, circuits can contain more than one current source provided that the circuit does not violate certain rules. One rule is that you must have a load across the current source. This is because a current source must maintain a specific level of current through it, and that current must go somewhere.

Another rule is that you cannot connect one current source directly in series with (in the path of) another current source. Refer to Figure 4-10.

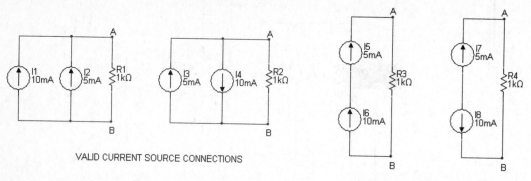

VALID CURRENT SOURCE CONNECTIONS

INVALID CURRENT SOURCE CONNECTIONS

Figure 4-10: Multiple Current Source Connections

The nature of a current source will explain why the two invalid current source connections are invalid. In both cases the 10 mA supply will try to keep the current through it at 10 mA, while the 5 mA supply will try to keep the current at 5 mA. It is clearly impossible to determine the current flowing between points A and B for ideal current sources, and for practical current sources these connections could damage the supplies. NEVER directly connect current sources in series. If you attempt to do so in a Multisim circuit then the software will warn you of the problem.

For the valid current source connections, called parallel connections, the current between points A and B will be the algebraic sum of the current sources. To understand this, consider what happens at the junction of the current sources. For the first circuit, *I1* will source 10 mA of current into the top junction and *I2* will source 5 mA. These currents will combine to form 10 mA + 5 mA = 15 mA that will flow through the resistor between points A and B. From point B the current will return to the current sources, with *I2* sinking 5 mA and *I1* sinking the remaining 10 mA. For the second circuit *I4* will source 10 mA into the bottom junction. *I3* will sink 5 mA of this and source it into the top junction, while the remaining 10 mA – 5 mA = 5 mA will flow from point B to point A through the resistor. This 5 mA will combine at the top junction with the 5 mA from *I3* to form 5 mA + 5 mA = 10 mA, which *I4* will sink.

To verify this, open the circuit Ex04-04, measure the current between points A and B, and record the results in Table 4-4. Remember to connect the meter so that the + terminal is closer to point A than the – terminal.

Table 4-4: Measured I_{AB} Values

	Circuit 1 (V)	Circuit 2 (V)
Calculated	15 mA	−5 mA
Measured		

4.4 Application Exercise

All voltages are measured with respect to some reference point. For circuits this reference point is called reference ground. As you will learn, the exact location of this reference point for measurement purposes is unimportant, although safety considerations may dictate using a specific point in the circuit. In this exercise you will see the effects of using different points as reference ground for circuit measurements.

1. Open practice circuit file Pr04-01 shown in Figure 4-11.

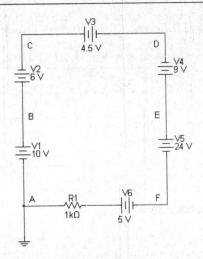

Figure 4-11: Reference Ground Practice Circuit

2. Verify that the voltage between point B and circuit ground (V_B) is +10.000 V. This value is already recorded in the appropriate cell of Table 4-5.

3. Repeat the measurements for points C through F and record the voltages to complete the top row of Table 4-5.

4. Move the circuit ground to points B through F in turn and record the circuit voltages to complete Table 4-5.

Table 4-5: Measured Circuit Voltages

		Measured Voltage (Positive Voltmeter Terminal) (V)					
		V_A	V_B	V_C	V_D	V_E	V_F
Ground Node (Negative Voltmeter Terminal) (V)	A	0	10.000				
	B		0				
	C			0			
	D				0		
	E					0	
	F						0

5. Use your measured values from Table 4-5 to complete Table 4-6. Recall that $V_{XY} = V_X - V_Y$ so that $V_{BA} = V_B - V_A$, $V_{CB} = V_C - V_B$, and so forth.

Table 4-6: Voltage Differences for Reference Ground Practice Circuit

		Measured Voltage (Positive Voltmeter Terminal) (V)					
		V_{BA}	V_{CB}	V_{DC}	V_{ED}	V_{FE}	V_{FA}
Ground Node (Negative Voltmeter Terminal) (V)	A						
	B						
	C						
	D						
	E						
	F						

Did the location of the reference ground affect the voltage differences in the circuit?

4.5 Section Summary

In this section you investigated the properties of ideal voltage and current sources. You also studied the concept of reference ground in a circuit and the effects of choosing different points as reference ground in a circuit. You should have found that the location of reference ground affects the voltages of points when measured relative to reference ground (Table 4-5), but not when measured relative to each other (Table 4-6). One important implication of this is that you should always measure the same voltage across a component or group of components regardless of the location of reference ground.

5. Introduction to Resistive Circuits

5.1 Introduction

All practical circuits possess resistance. A circuit that possesses only resistance is referred to as a resistive circuit. Resistance, as you have learned, is the opposition to current flow. Resistors allow you to add specific amounts of resistance to a circuit so that you can control the amount of voltage and current present in specific parts of the circuit. Although resistance opposes the flow of current, you may be surprised to find that adding resistance to a circuit can actually increase the current flowing in a circuit.

In previous sections you built and studied circuits that used fixed and variable resistors. Section 4 revealed some characteristics about resistors when you studied dc power sources. You saw, for example, that current would increase when the resistance in a circuit decreased, and that more voltage was necessary to push more current through a fixed value of resistance. Up to now, however, your experience with resistors has been mostly incidental to learning about using the Multisim software and other components.

In this section you will use the Multisim program to systematically study resistors. In particular, you will:

- Learn general characteristics of series and parallel resistive circuits.

- Study the relationship between resistance, voltage, and current.

- Learn how to copy and paste components in Multisim circuits.

5.2 Series Resistive Circuits

Most of the Multisim circuits you have built thus far have been series circuits, although the term "series" was not defined and seldom used. In practical terms, a series circuit is one for which only one current path exists. A logical conclusion from this is that the same current flows through all components in a series connection. In this section you will investigate the relationship between voltage, current, and resistance in a series resistive circuit.

5.2.1 Effect of Voltage on Current in Series Resistive Circuits

In this section you will measure the effect of changing voltage on current for a fixed resistance.

1. Open the circuit file Ex05-01 shown in Figure 5-1.

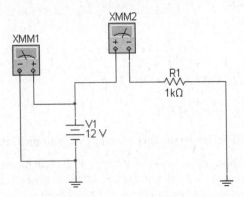

Figure 5-1: Series Resistance Current vs. Voltage Demonstration Circuit

2. Save the circuit as Ex05-01_Mod.

3. Open the component properties window for *VS* and change the voltage to the first value (1 V) in Table 5-1.

4. Click the **Run** switch to start the simulation.

5. Record the measured voltage and current values in the appropriate cells under the column "One resistor in series" in Table 5-1.

6. Click the **Run** switch to stop the simulation.

7. Repeat Steps 3 through 6 for the remaining values of *VS* shown in Table 5-1.

Next, modify the circuit to add a second 1 kΩ resistance in series with *R1*. Rather than change the value of *R1* to do this, you will copy *R1* to add a 1 kΩ resistor to the circuit.

1. If you are using Multisim 10, hold down the SHIFT key, click the ground symbol, and drag the ground symbol to the right far enough to insert another resistor to the right of *R1*. If you hold the SHIFT key down the cursor will move orthogonally (that is, straight up, down, left, or right—not diagonally). If you attempt to move the cursor diagonally, the cursor will "snap" in the orthogonal direction that the cursor has moved the farthest. Tap the RIGHT ARROW key once as needed to straighten the wiring connection between *R1* and the ground symbol.

    If you are using Multisim 9 software, select the ground symbol and press the RIGHT ARROW key repeatedly to move the ground symbol to the right. This method will also work in the Multisim 10 program.

2. Right-click *R1* to select *R1* and open the right-click menu.

3. Select **Copy** in the right-click menu.

4. Right-click away from *R1* to open the right-click menu.

5. Select **Paste** in the right-click menu. An outline image of the resistor copy will appear at the tip of the cursor.

6. Position the resistor copy on the wire to the right of *R1* and click to paste the copy into the circuit. Your circuit should now be similar to the circuit in Figure 5-2. Note that the Multisim program automatically renumbered the reference designator for the resistor copy, as components cannot have the same reference designator.

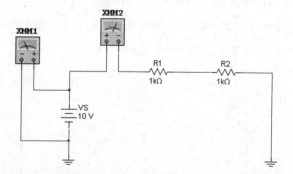

Figure 5-2: Modified Series Resistance Current vs. Voltage Demonstration Circuit

Now change the value of *VS* to 1 V, repeat the voltage and current measurement procedure you used for one series resistor, and record the voltage and current values in the cells under the column "Two resistors in series". Remember to stop the simulation when you are finished.

Finally, add a third 1 kΩ resistor in series with *R1* and *R2* and repeat the procedure to complete the cells under the column "Three resistors in series".

Table 5-1: Values for Current vs. Voltage for Series Resistors

VS	One resistor in series		Two resistors in series		Three resistors in series	
	Voltage (V)	Current (mA)	Voltage (V)	Current (mA)	Voltage (V)	Current (mA)
1 V						
2 V						
3 V						
4 V						
5 V						
6 V						
7 V						
8 V						
9 V						
10 V						

Use your data from Table 5-1 to plot the current vs. the voltage for each of the 3 circuit resistances in the graph below, labeling the plots as "1 resistor in series", "2 resistors in series", and "3 resistors in series".

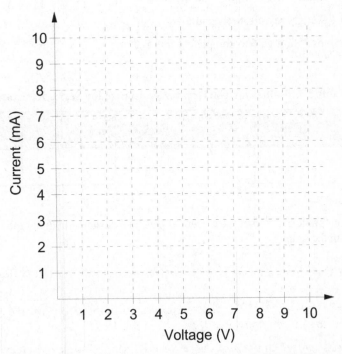

What conclusions about voltage and current can you draw from the plots? Can you draw any conclusion about resistance and current?

5.2.2 Effect of Resistance on Current in Series Resistive Circuits

From the graphs you drew you should have seen that current increases linearly with voltage, indicating that current is directly proportional to the voltage. You should also have noted that current decreased as you added resistors in series so that current appears to be inversely proportional to the number of series resistors. In the next experiment you will investigate this current vs. resistance relationship further.

1. Open the circuit Ex05-02 shown in Figure 5-3 .

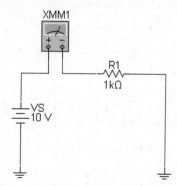

Figure 5-3: Series Resistance Current vs. Resistance Demonstration Circuit

2. Save the circuit as Ex05-02_Mod.

3. Click the **Run** switch to start the simulation.

4. Record the current for the circuit resistance.

5. Click the **Run** switch to stop the simulation.

6. Add a 1 kΩ series resistor to the circuit.

7. Repeat Steps 3 through 6 to complete Table 5-2.

Table 5-2: Values for Current vs. Resistance for Series Resistors

Number of 1 kΩ Resistors in Series	1	2	3	4	5	6	7	8	9	10
Measured Current (mA)										

Use your data from Table 5-2 to plot the current versus the resistance in the graph below. What conclusion can you draw from the graph?

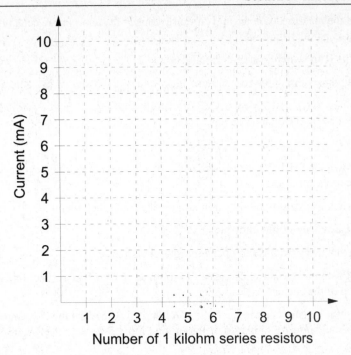

5.3 Parallel Resistive Circuits

In practical terms, a parallel circuit is one in which the components connect between the same two points. Since voltage is defined as the potential difference between two points, a logical conclusion is that parallel components have the same voltage across them. In this section you will investigate the relationship between voltage, current, and resistance in parallel circuits.

5.3.1 Effect of Voltage on Current in Parallel Resistive Circuits

In this section you will measure the effect of changing voltage on current for a fixed resistance.

1. Open the circuit Ex05-03 shown in Figure 5-4.

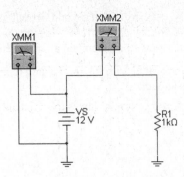

Figure 5-4: Parallel Resistance Current vs. Voltage Demonstration Circuit

2. Save the circuit as Ex05-03_Mod.

3. Open the component properties window for *VS* and change the voltage to the first value (1 V) in Table 5-3.

4. Click the **Run** switch to start the simulation.

5. Record the measured voltage and current values in the appropriate cells under the column "One resistor in parallel" in Table 5-3.

6. Click the **Run** switch to stop the simulation.

7. Repeat Steps 3 through 6 for the remaining values of *VS* shown in Table 5-3.

Next, modify the circuit to add a second 1 kΩ in parallel with *R1*. To add a parallel resistor, you will copy both *R1* and the ground to which it connects.

1. Click *R1* to select it.

2. Hold down the SHIFT key and click (SHIFT+click) on the ground symbol below R1 to add it to the selection.

3. Right-click away from the selection to open the right-click menu.

4. Select <u>C</u>opy in the right-click menu.

5. Right-click away from *R1* to open the right-click menu.

6. Select <u>P</u>aste in the right-click menu. An outline image of the resistor and ground copy will appear at the tip of the cursor.

7. Position the resistor copy to the right of *R1* and click to paste the copy into the circuit. Your circuit should now be similar to the circuit in Figure 5-5.

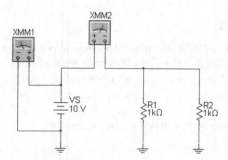

Figure 5-5: Parallel Resistance Modified Current vs. Voltage Demonstration Circuit

Now change the value of *VS* to 1 V, repeat the voltage and current measurement procedure you used for one parallel resistor, and record the voltage and current values in the cells under the column "Two resistors in parallel". Remember to stop the simulation when you are finished.

Finally, add a third 1 kΩ resistor in parallel with R1 and R2 and repeat the procedure to complete the cells under the column "Three resistors in parallel".

Table 5-3: Values for Current vs. Voltage for Parallel Resistors

VS	One resistor in parallel		Two resistors in parallel		Three resistors in parallel	
	Voltage (V)	Current (mA)	Voltage (V)	Current (mA)	Voltage (V)	Current (mA)
1 V						
2 V						
3 V						
4 V						
5 V						
6 V						
7 V						
8 V						
9 V						
10 V						

Use your data from Table 5-3 to plot the current versus the voltage for each of the three circuit resistances in the graph below, labeling the plots as "One resistor in parallel", "Two resistors in parallel", and "Three resistors in parallel".

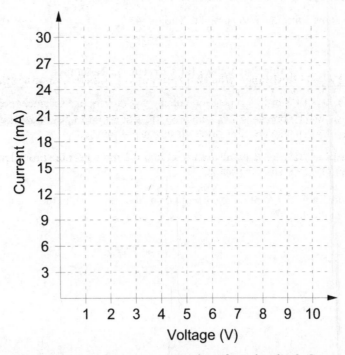

What conclusions about voltage and current can you draw from the plots? Can you draw any conclusion about resistance and current?

5.3.2 Effect of Resistance on Current in Parallel Resistive Circuits

From the graphs you drew you should have seen that current increases linearly with voltage, indicating that current is directly proportional to the voltage. You should also have noted that current increased as you added parallel resistance to the circuit so that current appears to be proportional to the number of parallel resistors. In the next experiment you will investigate this current vs. resistance relationship further.

1. Open the circuit Ex05-04 shown in Figure 5-6.

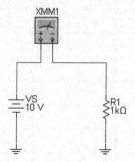

PARALLEL RESISTIVE CIRCUIT

Figure 5-6: Current vs. Resistance Demonstration Circuit

2. Save the circuit as Ex05-04_Mod.

3. Click the **Run** switch to start the simulation.

4. Record the current for the 1 kΩ circuit resistance.

5. Click the **Run** switch to stop the simulation.

6. Add a 1 kΩ parallel resistor to the circuit.

7. Repeat Steps 3 through 6 to complete Table 5-4.

Table 5-4: Values for Current vs. Resistance

Number of 1 kΩ Parallel Resistors	1	2	3	4	5	6	7	8	9	10
Measured Current (mA)										

Use your data from Table 5-4 to plot the current versus the resistance in the graph below. What conclusion can you draw from the graph?

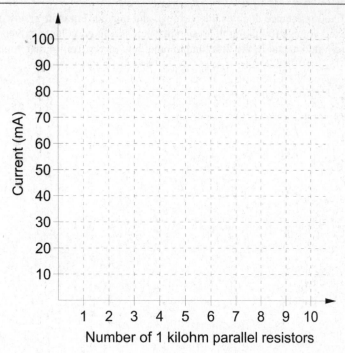

5.4 Application Exercise

In the experiment for Section 5.2.2 you added series 1 kΩ resistors to the circuit and measured the circuit current that resulted. You probably suspected that adding each resistor increased the total circuit resistance by 1 kΩ, so that the resistance of two resistors was 2 kΩ, that of three resistors was 3 kΩ, and so forth. To see whether this is so, open the circuit Ex05-02. Change the value of *R1* to each of the values shown in Table 5-5, and record the resulting current. How do the current values of Table 5-5 compare to your measurements in Table 5-2?

Table 5-5: Current for Increasing Series Resistance

Value of *R1*	Measured Current
1 kΩ	
2 kΩ	
3 kΩ	
4 kΩ	
5 kΩ	
6 kΩ	
7 kΩ	
8 kΩ	
9 kΩ	
10 kΩ	

5.5 Section Summary

In this section you studied the relationship between voltage, current, and resistance. Your measurements should have shown that current through a resistance is proportional to the applied voltage (doubling the voltage across a fixed resistance doubles the current), and that current is inversely proportional to the resistance for a fixed voltage (doubling the resistance across a fixed voltage halves the current through it). These findings are fundamental to the first fundamental law of electronics that you will study in the next section.

6. Ohm's Law

6.1 Introduction

The experiments you performed in Section 5 are similar to those performed in the 19th century by investigators who were studying the nature and properties of electricity. One of those early investigators was Georg Simon Ohm, a German physicist teaching mathematics in Köln (Cologne), Germany and for whom the units of resistance is named. His experiments, like yours, showed that current is directly proportional to the applied voltage and inversely proportional to the circuit resistance. Based on his studies, he presented an equation that related current, voltage, and resistance. Today we know this relationship as **Ohm's Law**:

$$I = V / R$$

where

I is the current in amperes,

V is the voltage in volts, and

R is the resistance in ohms.

You can use Ohm's Law to find any quantity if you know the other two quantities. For example, if you know the resistance and the voltage applied across it, you can determine the current through the resistance. There are actually three forms of Ohm's Law, depending upon the quantity you wish to find:

$$I = V / R$$

$$V = I \times R$$

$$R = V / I$$

You can use the three-variable "pie chart" shown in Figure 6-1 to represent these forms.

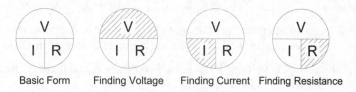

| Basic Form | Finding Voltage | Finding Current | Finding Resistance |

Figure 6-1: Pie Chart Representing Forms of Ohm's Law

To use the pie chart, start with the basic form. Then, simply cover the letter that represents the entity you wish to find. To find the formula for voltage, cover the section that contains the letter "V" to show "IR" (or "I × R") as the formula to find voltage. Similarly, to find the formula for resistance, cover the section that contains the letter "R" to show "V over I" (or "V / I").

In this section you will:

- Verify Ohm's law for resistive circuits.

- Learn to recognize basic faults from circuit measurements.

6.2 Verifying Ohm's Law

6.2.1 Verifying Ohm's Law for Circuit Resistance

1. Open the circuit file Ex06-01.

2. Run the simulation to determine the current for the circuit.

3. Record the measured current for the circuit, I_{MEAS}, under the column "Circuit 1" in Table 6-1.

4. Use Ohm's Law to calculate the resistance of the circuit from VS and I_{MEAS}. Enter this calculated resistance value, R_{CALC}, under the column "Circuit 1" in Table 6-1.

5. Disconnect the dc voltage source and measure the resistance, R_{MEAS}, between Point A and Point B. Record this value under the column "Circuit 1" in Table 6-1.

6. Calculate the difference between R_{MEAS} and R_{CALC} for the circuit. Record this value as R_{DIFF} under the column "Circuit 1" in Table 6-1.

7. Repeat Steps 1 through 6 for circuit files Ex06-02 through Ex06-04, using the columns "Circuit 2" through "Circuit 4" in Table 6-1, respectively.

Table 6-1: Resistance Calculations for Ohm's Law

Circuit Value	Circuit 1	Circuit 2	Circuit 3	Circuit 4
VS (V)				
I_{MEAS} (mA)				
R_{CALC} (kΩ)				
R_{MEAS} (kΩ)				
R_{DIFF} (kΩ)				

6.2.2 Verifying Ohm's Law for Circuit Voltage

1. Open the circuit file Ex06-05.

2. Record the value of $R1$ in Table 6-2.

3. Run the simulation to determine the current for the circuit.

4. Record the measured current for the circuit, I_{MEAS}, under the column "Circuit 1" in Table 6-2.

5. Use Ohm's Law to calculate the dc supply voltage of the circuit from $R1$ and I_{MEAS}. Enter this calculated resistance value, VS_{CALC}, under the column "Circuit 1" in Table 6-2.

6. Measure the dc supply voltage, VS_{MEAS}, between Point A and Point B. Record this value under the column "Circuit 1" in Table 6-2.

7. Calculate the difference between VS_{MEAS} and VS_{CALC} for the circuit. Record this value as VS_{DIFF} under the column "Circuit 1" in Table 6-2.

8. Repeat Steps 1 through 6 for circuit files Ex06-06 through Ex06-08, using the columns "Circuit 2" through "Circuit 4" in Table 6-2, respectively.

Table 6-2: Voltage Calculations for Ohm's Law

Circuit Value	Circuit 1	Circuit 2	Circuit 3	Circuit 4
$R1$ (kΩ)				
I_{MEAS} (mA)				
VS_{CALC} (V)				
VS_{MEAS} (V)				
VS_{DIFF} (V)				

6.2.3   Verifying Ohm's Law for Circuit Current

1. Open the circuit file Ex06-09.

2. Record the value of **R1** in Table 6-3.

3. Record the value of **VS** in Table 6-3.

4. Use Ohm's Law to calculate the circuit current I_{CALC} from **VS** and **R1**. Enter this calculated current value, I_{CALC}, under the column "Circuit 1" in Table 6-3.

5. Run the simulation.

6. Insert an ammeter between the dc voltage source and Point A and measure the circuit current, I_{MEAS}.

7. Record the measured current for the circuit, I_{MEAS}, under the column "Circuit 1" in Table 6-3.

8. Calculate the difference between I_{MEAS} and I_{CALC} for the circuit. Record this value as I_{DIFF} under the column "Circuit 1" in Table 6-1.

9. Repeat Steps 1 through 6 for circuit files Ex06-10 through Ex06-12, using the columns "Circuit 2" through "Circuit 4" in Table 6-3, respectively.

Table 6-3: Current Calculations for Ohm's Law

Circuit Value	Circuit 1	Circuit 2	Circuit 3	Circuit 4
R1 (kΩ)				
VS (V)				
I_{CALC} (mA)				
I_{MEAS} (mA)				
I_{DIFF} (mA)				

6.3   Application Exercise

Open the circuit file Pr06-01 shown in Figure 6-2. Use Ohm's Law to complete the missing circuit values in Figure 6-2. Then, use the Multisim software to modify the circuit values to each of those in Table 6-4 and simulate the circuit to verify that your calculated values are correct.

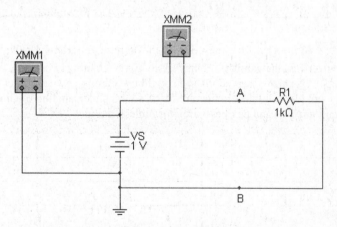

Figure 6-2: Circuit Value Calculation Circuit

Table 6-4: Calculated Circuit Values

VS	R1	I_{MEAS}
5.0 V	1.5 kΩ	
3.3 V		125 µA
	910 Ω	2.5 mA
12 V		13 mA
6.0 V	75 Ω	
	51 kΩ	500 µA
100 V	27 Ω	
48 V		750 µA
	680 Ω	145 mA

6.4 Troubleshooting Exercise

You can use Ohm's Law to determine whether a fault exists in a circuit. For example, if you know what the circuit values should be, you can calculate what the third value in a circuit should be. If the measured value does not match the calculated value then a fault must exist in the circuit.

Use Ohm's law to calculate I_{CALC} for each of the indicated circuits in Table 6-5. Then, use the Multisim software to open and simulate each circuit. Record the measured current as I_{MEAS} and determine whether the circuit contains a fault.

Table 6-5: Troubleshooting Circuit Values

Circuit Name	VS	R1	I_{CALC}	I_{MEAS}	Fault? (Yes/No)
T06-01	3.3 V	5.1 kΩ			
T06-02	5.0 V	3.3 kΩ			
T06-03	10 V	27 kΩ			

6.5 Section Summary

In this section you verified the validity of Ohm's Law for calculating resistance, voltage, and current in resistive dc circuits. You also observed that you could use Ohm's Law to predict the effects of changing circuit parameters and identify faults in the circuit. A sound understanding of Ohm's Law is fundamental to theoretical circuit analysis and practical circuit troubleshooting.

7.   Power in DC Circuits

7.1   Introduction

All electronic circuits require power to operate. Power is the rate at which energy is used (E / t), so power in a circuit will increase if

- the circuit uses more energy in the same amount of time,

- the circuit uses the same amount of energy faster, or

- the circuit uses more energy faster.

Electronic power is equal to the product of voltage and current and typically expressed in units of **watts**. The voltage (E / Q) represents the energy stored in the charges, and the current (Q / t) represents the rate at which the charges return to a lower energy state. A mathematical way to express this is

$$P = E / t = (E / Q) \times (Q / t) = V \times I$$

The most familiar form of the power equation for electronic circuits is $P = V \times I$, but there are 3 different forms depending upon the quantity you wish to find:

$$P = V \times I$$

$$V = P / I$$

$$I = P / V$$

As with Ohm's Law, you can use the three-variable "pie chart" in Figure 7-1 to represent these forms.

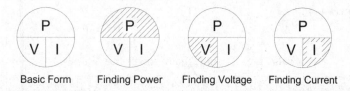

Figure 7-1: Pie Chart Representing Forms of Basic Power Equation

To use the pie chart, start with the basic form. Then, simply cover the letter that represents the entity you wish to find. To find the formula for voltage, cover the section that contains the letter "V" to show "P over I" (or "P / I") as the formula to find voltage. Similarly, cover the section that contains the letter "I" to show "P over V" (or "P / V") to find the formula for current.

You can combine the basic power equation and forms of Ohm's Law to express power directly in terms of circuit parameters of voltage, current, and resistance.

$$P = V \times I = (I \times R) \times I = (I \times I) \times R = I^2 \times R$$

$$P = V \times I = V \times (V / R) = (V \times V) / R = V^2 / R$$

In this section you will:

- Use the wattmeter to measure power in resistive dc circuits.

- Verify that power in equals power out in resistive dc circuits.

- Verify that calculated power equals measured power for resistive DC circuits.

7.2   The Wattmeter

One way to determine power in a circuit is to use a multimeter to measure circuit values and calculate power from one of the power formulas. Another way is to use the **wattmeter**, which is an instrument that measures the power in a circuit. Since power is the product of voltage and current, the wattmeter has two sets of inputs to measure both voltage and current simultaneously. The wattmeter uses these measurements to calculate the power. Figure 7-2 shows the minimized and expanded views of the Multisim wattmeter.

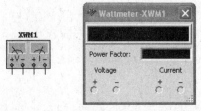

Figure 7-2: Multisim Wattmeter Views

To connect the wattmeter, you connect the voltage inputs (marked with a "V") to the circuit as you would connect a voltmeter and the current inputs (marked with an "I") to the circuit as you would connect an ammeter. The following example will illustrate this.

1. Open the circuit Ex07-01 shown in Figure 7-3. For this example you will measure the power in **R1**.

2. Save the circuit as Ex07-01_Mod.

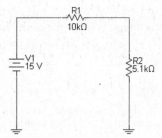

Figure 7-3: Wattmeter Demonstration Circuit

3. Select the Wattmeter tool from the Instruments toolbar and place it so that the voltage inputs are above **R1**, as shown in Figure 7-4.

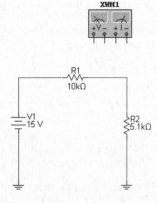

Figure 7-4: Placing the Wattmeter

4. Connect the wattmeter voltage inputs across *R1* as shown in Figure 7-5.

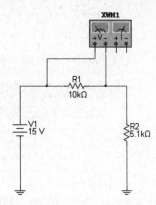

Figure 7-5: Connecting the Wattmeter Voltage Inputs

5. Next, you must connect the current inputs into the circuit so that they measure the current through *R1*. You can connect the ammeter in series with either the left or right side of *R1*. Since the current inputs are to the right of the voltage inputs it is easier to connect them into the circuit to the right side of *R1*. Select the wire segment between the junction to the right of *R1* and the top terminal of *R2*. Press the DELETE key to delete the selected segment. Your circuit should now look like that in Figure 7-6.

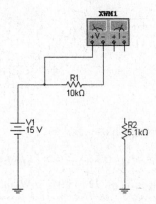

Figure 7-6: Opening the Circuit for the Current Input Connections

6. Connect the current inputs in series with the circuit between *R1* and *R2,* as shown in Figure 7-7.

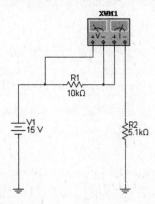

Figure 7-7: Connecting the Wattmeter Current Inputs

7. Double-click the wattmeter to open the expanded view.

8. Click the **Run** switch to run the simulation. After a few seconds the wattmeter should show the readings in Figure 7-8.

Figure 7-8: Wattmeter Power Measurement for *R1*

The wattmeter shows the values. The first value, in the large window, indicates that *R1* is dissipating 9.868 mW of power. The smaller window indicates that the power factor is 1.000. When you study power in ac circuits you will learn what "power factor" is. For now, just realize that the power factor of purely resistive circuits is always 1.

9. Click the **Run** switch to stop the simulation.

10. Connect the wattmeter as shown in Figure 7-9.

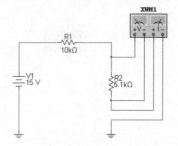

Figure 7-9: Wattmeter Connection

11. Verify that the power dissipation of *R2* is 5.033 mW and that the power factor is 1.000.

7.3 Power In vs. Power Out

For ideal circuits with 100% efficiency and no parasitic effects, the power that the energy source provides should equal the total power dissipated in the circuit. Follow the steps below to determine if this is true.

1. Open the circuit file Ex07-01.

2. Use a voltmeter to measure the value of *V1* and record this value in Table 7-1.

3. Use an ammeter between *V1* and *R1* to measure the total current, I_T, that the dc voltage source provides to the circuit and record this value in Table 7-1.

4. Enter your measured values P_{R1} and P_{R2} that you measured for *R1* and *R2* in Table 7-1.

5. Multiply *V1* and I_T to calculate the input power, P_{IN}, supplied by the dc voltage source and record this value in Table 7-1.

6. Add *P1* and *P2* to determine the total dissipated power, P_{OUT}, and enter this value in Table 7-1.

7. Calculate $P_{DIFF} = P_{IN} - P_{OUT}$ and enter this difference in Table 7-1. Is there a difference between the total input power and total dissipated power?

Table 7-1: Calculating the Difference Between Power In and Power Out

Parameter	Value
$V1$	
I_T	
$P1$	
$P2$	
$P_{IN} = V1 \times I_T$	
$P_{OUT} = P_{R1} + P_{R2}$	
$P_{DIFF} = P_{IN} - P_{OUT}$	

7.4   Application Exercises

7.4.1   Calculated Power from Voltages vs. Measured Power

1. Open the circuit file Pr07-01 shown in Figure 7-10.

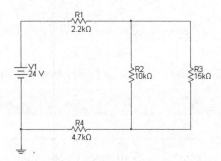

Figure 7-10: Calculated and Measured Power Practice Circuit 1

2. Measure the voltage for each component and enter the measured values as V_{MEAS} in Table 7-2.

3. Use the resistance and measured voltages to calculate the dissipated power P_{CALC} for each resistor and enter these values in Table 7-2.

4. Use the wattmeter to measure the power for each resistor and enter the measured values as P_{MEAS} in Table 7-2.

5. Indicate whether or not the measured power matches your calculated values for each resistor in Table 7-2.

Table 7-2: Calculated and Measured Power Practice Circuit 1 Values

Resistor	Value	V_{MEAS}	P_{CALC}	P_{MEAS}	$P_{CALC} = P_{MEAS}$?
$R1$	2.2 kΩ				
$R2$	10 kΩ				
$R3$	15 kΩ				
$R4$	4.7 kΩ				

7.4.2   Calculated Power from Currents vs. Measured Power

Open the circuit Pr07-02 as shown in Figure 7-11.

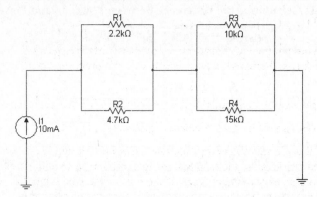

Figure 7-11: Calculated and Measured Power Practice Circuit 2

1. Measure the current for each resistor and enter the measured values as I_{MEAS} in Table 7-3.

2. Use the resistance and measured current values to calculate the dissipated power P_{CALC} for each resistor and enter these values in Table 7-3.

3. Use the wattmeter to measure the power for each resistor and enter the measured values as P_{MEAS} in Table 7-3.

4. Indicate whether or not the measured power matches your calculated values for each resistor in Table 7-3.

Table 7-3: Calculated and Measured Power Practice Circuit 2 Values

Resistor	Value	I_{MEAS}	P_{CALC}	P_{MEAS}	$P_{CALC} = P_{MEAS}$?
R1	2.2 kΩ				
R2	4.7 kΩ				
R3	10 kΩ				
R4	15 kΩ				

7.5   Section Summary

In this section you learned about the wattmeter, an instrument that measures power in a circuit, and verified the formulas for calculating power in a circuit. Power is a major practical consideration in many electronic applications. The power that a circuit must dissipate determines not only the expense of operating the circuit, but also the size, cost, expected operating life, and suitable applications for the circuit. Overheating due to excessive power dissipation is a primary cause of component failure in many circuits. When designing circuits a good rule of thumb is to select components whose power rating is at least double that of the calculated power dissipation. This helps to prolong component life and provide a safety margin for variations in operating temperature.

8.   Series Resistive Circuits

8.1   Introduction

Section 5 introduced some of the concepts of series resistive circuits. Series circuits are circuits for which only one current path exists. Series resistive circuits have two properties that are fundamental to circuit analysis.

1)   The total series resistance is equal to the sum of the individual series resistances.

2)   The current through each series resistance is identical.

The first property, which you investigated in Section 5.4, is fairly intuitive. As you add a certain amount of series resistance to a circuit it is only logical that the total resistance should increase by the same amount. The second property is also easy to understand. If there is only one path through which current can flow then it is only logical that the same current flows through all parts of the circuit.

In this section you will:

- Use the properties of series resistive circuits and Ohm's Law to analyze series resistive circuits.

- Use the Multisim software to verify your calculated circuit values.

- Learn to troubleshoot basic problems with series resistive circuits.

8.2   Pre-Lab

For each of the circuits shown in Figure 8-1, calculate the total series resistance and use Ohm's Law to calculate the total current and the voltage across each resistor. Record your calculations in Table 8-1.

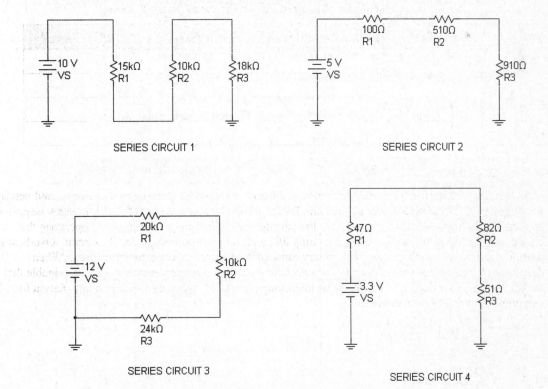

Figure 8-1: Example Series Circuits

Table 8-1: Calculated Circuit Values

Circuit Value	Circuit 1	Circuit 2	Circuit 3	Circuit 4
Source Voltage	10 V	5 V	12 V	3.3 V
Total Resistance				
Total Current				
$R1$	15 kΩ	100 Ω	20 kΩ	47 Ω
$R2$	10 kΩ	510 Ω	10 kΩ	82 Ω
$R3$	18 kΩ	910 Ω	24 kΩ	51 Ω
V_{R1}				
V_{R2}				
V_{R3}				

8.3 Design Verification

1. Use the Multisim software to open the circuit file Ex08-01.

2. Disconnect *VS* from the circuit and connect the ohmmeter between *R1* and ground.

3. Click the **Run** switch to start the simulation and measure the total series resistance. Record this value as "Total Resistance" under the column "Ex08-01" in Table 8-2.

4. Click the **Run** switch to stop the simulation and connect the ammeter between *VS* and *R1*.

5. Click the **Run** switch to start the simulation and measure the total current in the circuit. Record this value as "Total Current" under the column "Ex08-01" in Table 8-2.

6. Use the voltmeter to measure the voltage across each of the resistors *R1*, *R2*, and *R3*. Record these values as V_{R1}, V_{R2}, and V_{R3}, respectively, under the column "Ex08-01" in Table 8-2.

7. Click the **Run** switch to stop the simulation.

8. Open each of the circuits Ex08-02 through Ex08-04 in turn and repeat Steps 2 through 7 to complete Table 8-2. Do the values in Table 8-2 match those in Table 8-1?

Table 8-2: Measured Circuit Values

Circuit Value	Ex08-01	Ex08-02	Ex08-03	Ex08-04
Source Voltage	10 V	5 V	12 V	3.3 V
Total Resistance				
Total Current				
$R1$	15 kΩ	100 Ω	20 kΩ	47 Ω
$R2$	10 kΩ	510 Ω	10 kΩ	82 Ω
$R3$	18 kΩ	910 Ω	24 kΩ	51 Ω
V_{R1}				
V_{R2}				
V_{R3}				

8.4   Application Exercise

Although the Multisim program allows you to select any value of virtual resistor that you wish to place in a circuit, the real world is seldom so convenient. It is far more practical and cost-effective for companies to stock and use a limited number of resistor values for their circuits. Table 8-3 shows the standard 5% resistor values that are typically available for most designs.

Table 8-3: Table of Standard 5% Resistor Values

1.0	1.1	1.3	1.5	1.8	2.0	2.2	2.4
2.7	3.0	3.3	3.6	3.9	4.2	4.7	5.1
5.6	6.2	6.8	7.5	8.2	9.1		

Note that these standard resistor values represent the available numerical values and not the complete range of resistances. For example, the value "1.8" represents 5% resistance values of $0.18\ \Omega$, $1.8\ \Omega$, $18\ \Omega$, $180\ \Omega$, $1.8\ k\Omega$, $18\ k\Omega$, $180\ k\Omega$, $1.8\ M\Omega$, and $18\ M\Omega$. Similarly, "5.6" represents 5% resistance values of $0.56\ \Omega$, $5.6\ \Omega$, $56\ \Omega$, $560\ \Omega$, $5.6\ k\Omega$, $56\ k\Omega$, $560\ k\Omega$, $5.6\ M\Omega$, and $56\ M\Omega$.

As an exercise, design the series resistive circuit shown in Figure 8-2 using the above standard values. The circuit must nominally draw 10 μA from a 9 V dc voltage source, use only three resistors, and drop the voltages shown across each of the resistors. Record the values of your chosen resistors and the calculated values for the total current, the voltage drop across each resistor, and the percent error between your calculated and ideal values in Table 8-4. Use the formula % Error = [(Calculated – Ideal) / Ideal] × 100% to calculate the percent error. Finally, use the Multisim program to verify your design values and calculations.

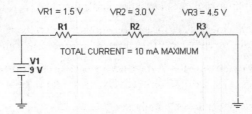

Figure 8-2: Application Exercise Circuit 1

Table 8-4: Design Calculations for Circuit 1

Circuit Value	Ideal	Calculated	% Error
Source Voltage	9 V	9 V	0%
Total Current	10 μA		
R1			
R2			
R3			
VR1	1.5 V		
VR2	3.0 V		
VR3	4.5 V		

After some testing, the Engineering Manager decides to change the design requirements for your circuit to those of Figure 8-3. Repeat the above design exercise to complete Table 8-5 and use the Multisim software to verify your design values and calculations.

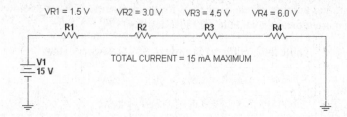

Figure 8-3: Application Exercise Circuit 2

Table 8-5: Design Calculations for Circuit 1

Circuit Value	Ideal	Calculated	% Error
Source Voltage	15 V	15 V	0%
Total Current	15 mA		
R1			
R2			
R3			
R4			
VR1	1.5 V		
VR2	3.0 V		
VR3	4.5 V		
VR4	6.0 V		

8.5   Troubleshooting Exercise

The most common faults that arise in resistive circuits are open circuits and short circuits. When a component short circuits, the current bypasses the component resistance (the current takes a "shortcut" around the opposition to current flow) so that the effective resistance becomes 0 Ω. When a component opens its resistance becomes effectively infinite so that no current can flow. Both these faults are relatively easy to identify and isolate.

- When a short circuit occurs in a series resistive circuit the total resistance decreases so that, by Ohm's Law, the total current will increase and be greater than the calculated value for the circuit. The resistance of the shorted component will be 0 Ω so that, again by Ohm's Law, the voltage drop across the resistor will be 0 V.

- When a resistor opens in a series resistive circuit current is unable to flow so that the total current is 0 A. The voltage drop across the open resistor will be equal to the source voltage.

A somewhat more difficult fault to isolate is a shifted resistance value. This fault can occur when environmental factors damage or age the resistor. The fault is not serious enough to open or short the component, but will change its value from the nominal, or stated, value. One way to identify the shifted value is to measure the total current and use Ohm's Law to determine the expected voltage drop across each resistor. The failed resistor is the one whose voltage drop does not match the calculated value. Another way is to measure the voltage across each resistor and calculate V / R for each one. The resistor whose V / R calculation does not match the others is the failed resistor.

Open the troubleshooting circuits Tr08-01 through Tr08-04 and complete Table 8-6. Based on your calculated and measured circuit values, indicate which component, if any, has a fault and the nature of the fault (open, short, or shifted value). If you believe no fault exists, indicate "None".

Table 8-6: Troubleshooting Circuit Values

Circuit Value	Tr08-01	Tr08-02	Tr08-03	Tr08-04
Source Voltage	10 V	5 V	12 V	3.3 V
Calculated Total Resistance				
Calculated Total Current				
Measured Total Current				
R1	15 kΩ	100 Ω	20 kΩ	47 Ω
R2	10 kΩ	510 Ω	10 kΩ	82 Ω
R3	18 kΩ	910 Ω	24 kΩ	51 Ω
Calculated V_{R1}				
Measured V_{R1}				
Calculated V_{R2}				
Measured V_{R2}				
Calculated V_{R3}				
Measured V_{R3}				
Suspected Circuit Fault (if any)				

8.6   Section Summary

In this section you used Ohm's Law and some principal circuit characteristics to analyze series resistive circuits and design series circuits to meet specific design requirements. You also learned how to identify common faults in series resistive circuits and used circuit measurements in the Multisim program to locate and identify faults in series resistive circuits. In the next section you will apply another fundamental law of electronics to analyze series resistive circuits.

9. Kirchhoff's Voltage Law

9.1 Introduction

In 1845 a German physicist named Gustav Richard Kirchhoff formulated two fundamental laws of electronics that are today named for him. These laws, with Ohm's Law, form the basis of virtually all circuit analysis that you will perform. This section will cover the law known as Kirchhoff's Voltage Law, sometimes abbreviated KVL. You will study the other law, known as Kirchhoff's Current Law, in another section.

Kirchhoff's Voltage Law states that the algebraic sum of voltages around any closed loop is zero. This means that if you trace any path in a circuit so that you end up at the same point from which you began, the voltage rises will equal the voltage drops. An analogy to this would be that of hiking in the mountains. No matter which path you may take, you will always end up at the same altitude at which you began and have the same potential energy with which you began when you return to your starting point, even if your legs say otherwise. Kirchhoff's Voltage Law is useful in its own right for analyzing simple circuits, but also forms the basis of loop current analysis that you will use later to analyze more complicated networks.

Voltage Polarity

An "algebraic sum" means that you must consider the sign as well as the magnitude of the quantities you are adding. Up to now you have been concerned only with the magnitude of voltages. Kirchhoff's Voltage Law requires you to consider the polarity of voltages as well. All voltages have polarities, indicated by "+" and "–" signs, which indicate the relative energy of the two points across which the voltage is measured. The standard convention indicates that the positive side has more energy than the negative side and that current flows from the positive to negative terminal of a dc source. Charges at the positive terminal of a battery have more energy than charges at the negative terminal, because the battery supplies energy to the charges as they move from the negative to positive terminal. Conversely, charges entering a resistor have more energy than charges leaving a resistor, as the resistor dissipates energy as power in the form of heat. Consequently the voltages across resistors are positive at the end at which current enters relative to the end from which current exits. Refer to Figure 9-1.

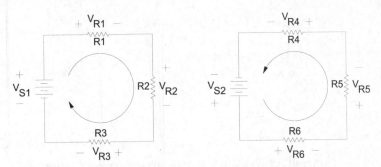

Figure 9-1: Voltage Polarities in Series Resistive Circuit

For the dc voltage source V_{S1}, the current will flow clockwise as shown, and develop voltages V_{R1}, V_{R2}, and V_{R3} with the indicated polarities across resistors $R1$, $R2$, and $R3$. Similarly, for the dc voltage source V_{S2}, the current will flow counterclockwise as shown, and develop voltages V_{R4}, V_{R5}, and V_{R6} with the indicated polarities across resistors $R4$, $R5$, and $R6$. The polarity of the dc voltmeter must match the polarity of the component to show a positive voltage reading. Similarly, current must flow into the + terminal of the dc ammeter for a positive current reading. A negative reading means that the polarity of the meter terminals is opposite to the actual voltage polarity or direction of current flow.

In this section you will:

- Verify the validity of Kirchhoff's Voltage Law.

- Use Kirchhoff's Voltage Law to analyze a circuit.

9.2   Pre-Lab

Use Kirchhoff's Voltage Law and Ohm's Law to determine the circuit values for the circuit in Figure 9-2 and record your calculated values in the "Calculated" column of Table 9-1. Initially assume that current flows in a clockwise direction. Indicate the actual direction of current flow based on the sign of your calculated result for total current. From Kirchhoff's Voltage Law

$$VS1 - V_{R1} - V_{R2} + VS2 - V_{R3} - VS3 - V_{R4} - V_{R5} - VS4 - V_{R6} = 0$$

Note that the values of *VS3* and *VS4* are negative as current enters the positive terminal and leaves the negative terminal, representing a voltage drop.

From Ohm's Law the voltage across each resistor is the resistor value times the total current I_T so

$$VS1 - (I_T)(R1) - (I_T)(R2) + VS2 - (I_T)(R3) - VS3 - (I_T)(R4) - (I_T)(R5) - VS4 - (I_T)(R5) = 0$$

Combining terms gives

$$VS1 + VS2 - VS3 - VS4 - I_T\,(R1 + R2 + R3 + R4 + R5 + R6) = 0$$

From this

$$VS1 + VS2 - VS3 - VS4 = I_T\,(R1 + R2 + R3 + R4 + R5 + R6)$$

so that

$$I_T = (VS1 + VS2 - VS3 - VS4)\,/\,(R1 + R2 + R3 + R4 + R5 + R6)$$

By substituting in the voltage source and resistance values you can determine the value of I_T. You can then use Ohm's Law to determine the voltage across each resistor for the value of I_T and the value of each resistor.

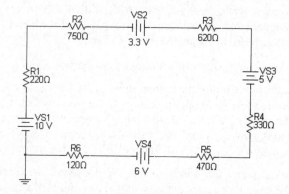

Figure 9-2: Kirchhoff's Voltage Law Exercise Circuit

9.3   Design Verification

Open the circuit file Ex09-01. Complete the values for the "Measured" column of Table 9-1. Use the polarity of the dc ammeter to determine the actual direction of current flow in the circuit.

Table 9-1: Calculated and Measured Circuit Values for Figure 9-2

Circuit Value	Calculated	Measured
I_T		
V_{R1}		
V_{R2}		
V_{R3}		
V_{R4}		
V_{R5}		
V_{R6}		
Current Flow Direction		

9.4 Application Exercise

In some cases, such as in those in Figure 9-1, the actual direction of current flow and voltage polarities are easy to determine. For other cases these may be less obvious. Consider the circuit in Figure 9-3.

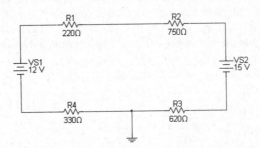

Figure 9-3: Kirchhoff's Voltage Law Demonstration Circuit

In these cases *assume* a direction for the current flow. If the current flows out of the positive terminal of a dc voltage source the dc source is supplying energy to the circuit (discharging) and is a voltage rise. If current flows into the positive terminal of a dc voltage source the dc source is drawing energy from the circuit (charging) and is a voltage drop. All resistors draw energy from the circuit and are voltage drops. All voltage rises are positive values and all voltage drops are negative values.

To see how this works, assume that current is flowing clockwise. Start from the ground and work your way clockwise around the circuit. You can actually start at any point but ground is a convenient reference point.

1) The voltage across *R4* (V_{R4}) is a voltage drop and is negative.

2) A clockwise current would flow out of the positive terminal of *VS1*, so that *VS1* is a voltage rise and positive.

3) The voltage across *R1* (V_{R1}) is a voltage drop and is negative.

4) The voltage across *R2* (V_{R2}) is a voltage drop and is negative.

5) A clockwise current would flow into the positive terminal of *VS2*, so that *VS2* is a voltage drop and negative.

6) The voltage across *R3* (V_{R3}) is a voltage drop and is negative.

Kirchhoff's Voltage Law requires that all these voltage must sum to 0 so that

$$(-V_{R4}) + VS1 + (-V_{R2}) + (-V_{R3}) + (-VS2) + (-V_{R3}) = 0$$

From Ohm's Law, the voltages across each resistor is just the total current times the resistor value, so that

$$(-I_T \times 330\ \Omega) + 12\ V + (-I_T \times 220\ \Omega) + (-I_T \times 750\ \Omega) + (-15\ V) + (-I_T \times 620\ \Omega) = 0$$

Collecting terms, rearranging, and simplifying yields

$$12\ V + (-15V) + (-I_T) \times (330\ \Omega + 220\ \Omega + 750\ \Omega + 620\ \Omega) = 0$$

$$-3\ V + (-I_T) \times (1920\ \Omega) = 0$$

Solving for I_T then gives

$$(-I_T) = 3\ V / (1920\ \Omega)$$

$$I_T = -3\ V / (1920\ \Omega) = -1.56\ mA$$

The calculated answer is a negative value, which means that the assumed direction of current flow is incorrect. The current in the circuit actually flows in a counterclockwise direction. In retrospect this seems logical, as the 15V source has a higher potential than the 12V supply. Open the circuit Ex09-02 and determine the value and direction of current flow for the circuit.

9.5   Section Summary

In this section you were introduced to Kirchhoff's Voltage Law and its use in analyzing series resistive circuits, especially those with more than one dc voltage source. You also learned the significance of voltage polarities, current direction, and signs of calculated circuit values. In a later section these concepts will be important when you use techniques such as loop current analysis to analyze circuit networks.

10. Voltage Dividers

10.1 Introduction

10.1.1 Basic Voltage Dividers

In Section 8.4 you designed two series resistive circuits that developed specific voltages across their component resistors. These circuits were examples of **voltage dividers**. As the name suggests, the purpose of a voltage divider is to divide down an applied voltage so that the circuit provides a specific fraction of the applied voltage. Some voltage dividers, like those you designed, provide multiple fractional voltages. Voltage dividers are a very common type of circuit. Some common applications are providing feedback in amplifier circuits, biasing voltages in transistor circuits, and proportional voltages for node voltage checks.

Voltage dividers are so common that there is a special equation to calculate the voltage output(s) of a voltage divider. You can derive this equation from Ohm's Law. Refer to Figure 10-1, which shows a series resistive circuit that consists of *n* resistors.

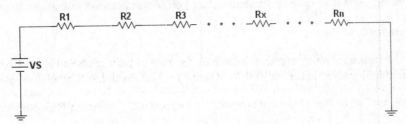

Figure 10-1: Voltage Divider Circuit

From Ohm's Law, the voltage Vx across any resistor Rx is equal to $I_T \times Rx$, where I_T is the total current. Again from Ohm's Law, the total current I_T is equal to VS / R_T, where R_T is the total resistance. Therefore

$$Vx = I_T \times Rx = (VS / R_T) \times Rx$$

$$= VS \times (Rx / R_T)$$

This formula can provide some valuable clues for troubleshooting. If a resistor (other than Rx) opens, then R_T becomes infinite and $Vx = 0$ V. If Rx opens, then $Vx = VS$. If a resistor (other than Rx) shorts, then R_T decreases and both I_T and Vx increase. If Rx shorts, then $Vx = 0$ V just as if another resistor opened. To tell whether Rx shorted or another resistor opened, measure the voltage across another resistor. If the other voltage is not 0 V, then Rx is shorted.

You can rearrange this equation to find the resistance value Rx that you require to develop a specific voltage Vx from an applied voltage VS:

$$Rx = (Vx / VS) \times R_T$$

Note that while this is useful for two-resistor voltage dividers, it is less useful for voltage dividers that consist of three or more resistors unless the resistors are of equal value, or some mathematical relationship exists between them so that you can express R_T directly in terms of Rx.

10.1.2 Multi-Tap Voltage Dividers

Many voltage dividers are multi-tap voltage dividers, which provide multiple voltages from several access points, or "taps," as in Figure 10-2. The main difference with these voltage dividers is that more than one resistor develops the output voltage so that Rx is not always the resistance of a single resistor. For Figure 10-2, Rx for *VOUT1* is $R2 + R3 + R4$, Rx for *VOUT2* is $R3 + R4$, and Rx for *VOUT3* is *R4*.

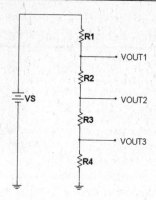

Figure 10-2: Multi-Tap Voltage Divider

Troubleshooting a multi-tap resistor is similar to that of a general voltage divider, except that **Rx** may consist of more than one resistor. In this case, if any of the resistors that make up **Rx** opens then **Vx = VS**. If any of the resistors that make up **Rx** shorts, then **VOUT** will decrease rather measure 0 V.

10.2 Pre-Lab

Calculate the resistance **Rx** and total resistance **R$_T$** for each of the circuits in Figure 10-3, and then use the voltage divider formula to calculate **VOUT**. Record your results in the "Calculated" column of Table 10-1.

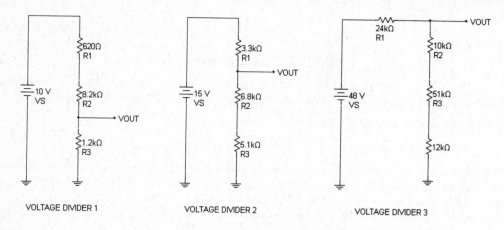

Figure 10-3: Voltage Divider Circuits

10.3 Design Verification

Open the circuit file Ex10-01.

1. Click the **Run** switch to start the simulation.

2. Measure the voltage between **VOUT** and ground.

3. Record your measured voltage as **VOUT** in the "Measured" column of "Voltage Divider 1" in Table 10-1.

4. Click the **Run** switch to stop the simulation.

Repeat Steps 1 through 4 for the circuit files Ex10-02 and Ex10-03 and record the values of **VOUT** in the "Measured" column under "Voltage Divider 2" and "Voltage Divider 3", respectively, in Table 10-1.

Table 10-1: Voltage Divider Calculated and Measured Values

Circuit Value	Voltage Divider 1		Voltage Divider 2		Voltage Divider 3	
	Calculated	Measured	Calculated	Measured	Calculated	Measured
VS	10 V		15 V		48 V	
Rx						
R_T						
VOUT						

10.4 Application Exercise

One design factor in voltage dividers is power dissipation. Because the resistors in the voltage dividers always draw current from the voltage source they are always dissipating power according to the equation

$$P_T = VS^2 / R_T$$

where P_T is the total dissipated power, VS is the applied voltage, and R_T is the total resistance of the voltage divider. One solution to this is to increase the value of R_T while keeping the resistor ratios the same. Consider the voltage divider circuits in Figure 10-4.

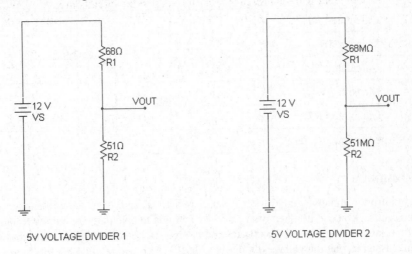

Figure 10-4: Increasing Voltage Divider Resistance

For each of the circuits, calculate the total resistance, total current, total power, and *VOUT*. Record these values in Table 10-2.

Table 10-2: Voltage Divider Values for Figure 10-4

Circuit Value	Voltage Divider 1	Voltage Divider 2
VS	12 V	12 V
RT		
IT		
PT		
Calculated *VOUT*		
Measured *VOUT*		

Open the circuit file Pr10-01, run the circuit simulation, and record the measured *VOUT* voltage as "Measured *VOUT*" under the column "Voltage Divider 1". Then open the circuit Pr10-02, run the circuit simulation, and record the measured *VOUT* voltage as "Measured *VOUT*" under the column "Voltage Divider 2".

What is the tradeoff for increasing voltage divider resistance to reduce the power consumption?

10.5 Troubleshooting Exercise

Open each of the circuits Tr10-01 through Tr10-03 and complete Table 10-3 with your measured and calculated values. From your values in Table 10-3, indicate whether or not a fault exists in each circuit and, if so, the nature of the fault.

Table 10-3: Troubleshooting Circuit Values

Circuit Value	Tr10-01	Tr10-02	Tr10-03
Measured *VS*			
Measured *VOUT*			
Measured I_T			
Measured V_{R1}			
Calculated *R1*			
Measured V_{R2}			
Calculated *R2*			
Measured V_{R3}			
Calculated *R3*			
Suspected Fault (if any)			

10.6 Section Summary

This section introduced the voltage divider, a series resistive circuit that you will encounter in many other electronic circuits. You can use the voltage divider formula to calculate the output voltage directly from the applied voltage, the total series resistance, and the resistance across which the output voltage is taken. This formula assumes that the voltage divider is a "stiff" one, meaning that the load across the output draws negligible current compared to the current drawn by the voltage divider itself. In the practice exercise, you discovered that very large voltage divider resistances would draw less current and dissipate less power, but that the load across the voltage divider could affect the measured output voltage. In a future section you will learn more about loading effects and the criterion for ensuring that the measured output and calculated output are acceptably close.

11. Parallel Resistive Circuits

11.1 Introduction

Section 5 introduced some of the concepts of parallel resistive circuits. Parallel circuits are circuits for which components have the same applied voltage. As you have seen, adding resistors in parallel increases the total current. Parallel resistive circuits have two properties that are fundamental to circuit analysis.

1) The total parallel resistance is equal to the reciprocal of the sum of the reciprocals of parallel resistances, or $R_T = 1 / (1/R1 + 1/R2 + ... + 1/Rn)$.

2) The voltage across each parallel resistance is identical.

The first property follows from the definition of a parallel circuit and Ohm's Law. Consider the circuit in Figure 11-1.

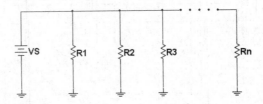

PARALLEL RESISTIVE CIRCUIT

Figure 11-1: Parallel Resistive Circuit

Because *VS* is across each of the resistors, Ohm's Law dictates that the currents through the resistors, called *branch currents*, are

$$I_{R1} = VS / R1, \; I_{R2} = VS / R2, \; ... \; I_{Rn} = VS / Rn.$$

The resistors draw these currents from the dc voltage source so that the total current is the sum of the individual currents. This gives

$$I_T \;\; = I_{R1} + I_{R2} + ... + I_{Rn} = VS / R1 + VS / R2 + ... + VS / Rn$$
$$= VS \times (1/R1 + 1/R2 + ... + 1/Rn)$$

But from Ohm's Law $R_T = VS/I_T$, so

$$R_T \;\; = VS / [VS \times (1/R1 + 1/R2 + ... + 1/Rn)] = (VS / VS) \times [1 / (1/R1 + 1/R2 + ... + 1/Rn)]$$
$$= 1 / (1/R1 + 1/R2 + ... + 1/Rn)$$

Note that if the resistor values are identical so that $R1 = R2 = ... = Rn = R$, then R_T reduces to $1 / (n \times 1/R)$ $= 1 / (n / R) = R / n$.

The second property, that the voltage across each parallel resistor is identical, follows directly from the definition of a parallel circuit.

In this section you will:

- Learn to work with groups of components in Multisim.

- Use the properties of parallel resistive circuits and Ohm's Law to analyze parallel resistive circuits.

- Use the Multisim software to verify your calculated circuit values.

- Learn to troubleshoot basic problems with parallel resistive circuits.

11.2 Pre-Lab

For the circuit shown in Figure 11-2, calculate the branch currents, total current, and total resistance. Enter these values under the "Calculated" columns for the appropriate parallel resistive circuit in Table 11-1.

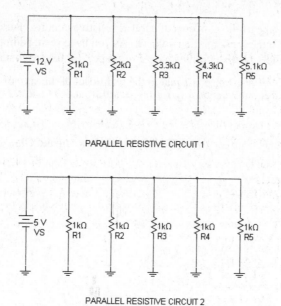

PARALLEL RESISTIVE CIRCUIT 1

PARALLEL RESISTIVE CIRCUIT 2

Figure 11-2: Parallel Resistive Circuits

11.3 Design Verification

Almost any operation that you can perform on a single component you can also perform on a group of two or more components. This is very convenient if you will be working on circuits, such as parallel circuits, that consist of very similar sections. You will use these methods to construct the circuits shown in Figure 11-2.

1. Open the circuit file Ex11-01 shown in Figure 11-3.

2. Save your circuit as Ex11-01_Mod.

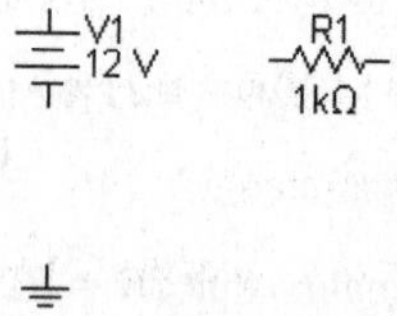

Figure 11-3: Parallel Resistive Circuit 1 Starting Circuit

3. Right-click *R1* to select it and open the right-click menu.

4. Select **Copy** from the right-click menu.

5. Right-click away from *R1* to open the right-click menu.

6. Select **Paste** from the right-click menu.

7. Click to the right of *R1* to place the copy of the resistor. The Multisim program will automatically number it as *R2*. Your circuit should look like that in Figure 11-4.

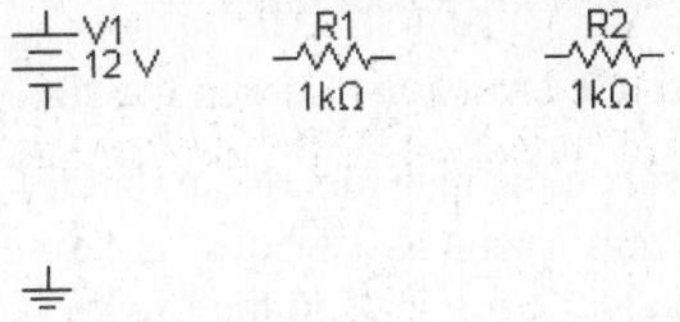

Figure 11-4: Parallel Resistive Circuit 1 with Two Resistors

8. Select *R1* and *R2*. You could do this by selecting *R1* and then holding down the SHIFT key and clicking on *R2*, but a faster way is to **window** around *R1* and *R2*. To do this, click above and to the left of *R1*, hold down the mouse key, and drag the cursor down and to the right. As you do so a rectangle will appear. When the rectangle encloses both *R1* and *R2*, release the mouse button. Dashed blue boxes will appear around *R1* and *R2*, indicating that both are selected.

    Windowing is a faster way to select multiple objects than the SHIFT + click method, but not always as precise. If you accidentally select a component that you did not intend to add to the group, you can SHIFT + click the component you wish to unselect. Note that you can use this method to selectively remove objects from groups you created using the SHIFT + click method as well. SHIFT + click will add an unselected object to a group, and remove a selected object from a group.

9. Right-click to open the right-click menu.

10. Select **Copy** from the right-click menu.

11. Right-click away from *R1* and *R2* to open the right-click menu.

12. Select **Paste** from the right-click menu.

13. Click to the right of *R2* to paste the copy of the two resistors. The Multisim program will automatically number them as *R3* and *R4*. Your circuit should now look like that of Figure 11-5.

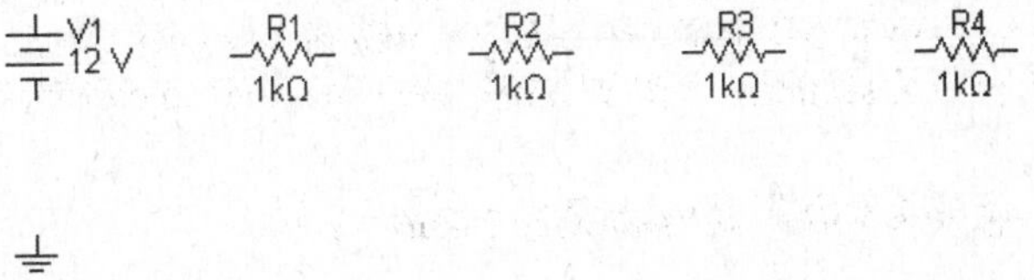

Figure 11-5: Parallel Resistive Circuit 1 with Four Resistors

14. Right-click *R4* to select it and open the right-click menu.

15. Select **Copy** from the right-click menu.

16. Right-click away from *R4* to open the right-click menu.

17. Select **Paste** from the right-click menu.

18. Click to the right of *R4* to place the final resistor. The Multisim program will automatically number it as *R5*. Your circuit should now look like that of Figure 11-6.

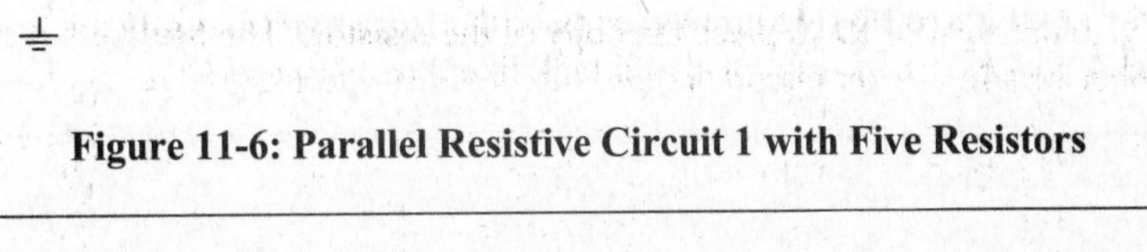

Figure 11-6: Parallel Resistive Circuit 1 with Five Resistors

19. Window around **R1** through **R5** to select them.

20. Right-click one of the selected resistors to open the right-click menu.

21. Select **90 Clockwise** to rotate the resistors. Note that this individually rotates each of the resistors, rather than rotating the entire group as a whole. Holding down the Control key and pressing "R" (CTRL + R) will also rotate the resistors in the group.

22. With the resistors still selected, click on and drag one of the resistors. As you do so you will move the entire group. Move the group up so that the resistors are approximately in-line with the dc voltage source. Refer to Figure 11-7.

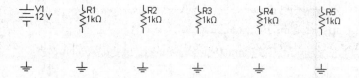

Figure 11-7: Parallel Resistive Circuit 1 with Rotated Resistors

23. Copy the ground symbol under the dc voltage source so that there is a ground symbol under each resistor, as shown in Figure 11-8.

Figure 11-8: Parallel Resistive Circuit 1 with Resistor and Ground Symbols Placed

24. Connect the dc voltage source, resistors, and ground symbols as shown in Figure 11-9.

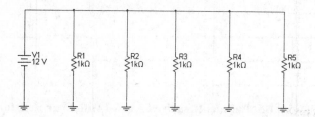

Figure 11-9: Parallel Resistive Circuit 1 with Components Placed and Wired

25. Use the Properties window for each component to change the dc voltage source reference designator to **VS** and the resistor values to match those of Figure 11-10.

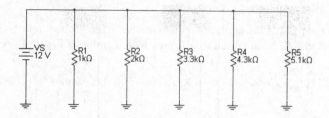

Figure 11-10: Completed Parallel Resistive Circuit 1

26. Save your circuit to keep your work to this point. This completes Parallel Resistive Circuit 1.

27. Use the Properties window for the dc voltage source and each resistor to change the value of the dc voltage source to 5 V and each of the resistors to 1 kΩ, as shown in Figure 11-11.

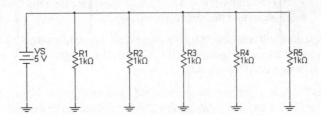

Figure 11-11: Parallel Resistive Circuit 2

28. Save your modified circuit as circuit file Ex11-02.

Now that you have created the parallel resistive circuits, simulate each circuit and complete Table 11-1 by recording the branch currents, total currents, and total resistance. Determine the total resistance by disconnecting the dc voltage source and using the ohmmeter to measure the total resistance directly.

Table 11-1: Parallel Resistive Circuit Values

Circuit Value	Parallel Resistive Circuit 1		Parallel Resistive Circuit 2	
	Calculated	Measured	Calculated	Measured
I_{R1}				
I_{R2}				
I_{R3}				
I_{R4}				
I_{R5}				
I_T				
R_T				

11.4 Application Exercise

Consider the circuit of Figure 11-12.

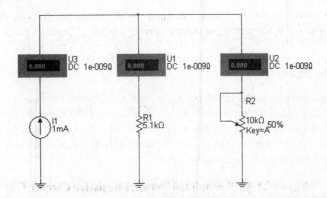

Figure 11-12: Parallel Resistive Circuit with DC Current Source

Because the voltage across parallel resistors is equal, then from Ohm's Law

$$V_{R1} = V_{R2} \rightarrow I_{R1} \times R1 = I_{R2} \times R2$$

Rearranging gives

$$I_{R1} \times R1 = I_{R2} \times R2 \rightarrow I_{R1} / I_{R2} = R2 / R1$$

This means that the current through the branches of a parallel resistive circuit should be inversely proportional to the resistance of the branches. If the value of **R2** is twice the value of **R1**, then the current through **R1** should be twice that through **R2**.

Open the circuit file Pr11-01. Increase the value of **R2** by pressing the "A" key and decrease the value of **R2** holding down the SHIFT key and pressing the "A" key. Record the value of total current and branch currents in Table 11-2.

Table 11-2: Current Source Parallel Resistive Circuit Measurements

R2 Setting	I_T (mA)	I_{R1} (mA)	I_{R2} (mA)	R2 Setting	I_T (mA)	I_{R1} (mA)	I_{R2} (mA)
0% (0 kΩ)				60% (6 kΩ)			
10% (1 kΩ)				70% (7 kΩ)			
20% (2 kΩ)				80% (8 kΩ)			
30% (3 kΩ)				90% (9 kΩ)			
40% (4 kΩ)				100% (10 kΩ)			
50% (5 kΩ)							

Use the data from Table 11-2 to complete Table 11-3.

Table 11-3: Parallel Resistive Circuit Branch Current Relationship

R2 Setting	R2 / R1	I_{R1}/I_{R2}	R2 Setting	R2 / R1	I_{R1}/I_{R2}
0%			60%		
10%			70%		
20%			80%		
30%			90%		
40%			100%		
50%					

Are the branch currents inversely proportional to the branch resistances?

11.5 Troubleshooting Exercise

Because the voltage across parallel resistors is identical, a voltmeter is typically not useful for isolating single-point faults in a parallel resistive circuit. Typically you must measure the branch currents and/or branch resistances to isolate the fault. One exception is if one branch opens and the current through each branch is different. If this happens then the difference between the measured current and calculated current will be the calculated current for the open branch.

Open each of the troubleshooting circuits Tr11-01 through Tr11-03 and complete Table 11-4. Based on your measured values, identify the suspected fault, if any.

Table 11-4: Troubleshooting Circuit Measured Values

Circuit Value	Tr11-01	Tr11-02	Tr11-03
I_T			
I_{R1}			
I_{R2}			
I_{R3}			
I_{R4}			
I_{R5}			
Suspected Fault (if any)			

11.6 Section Summary

This section demonstrated some techniques for working in Multisim with groups of components. These techniques are useful if you need to create or manipulate similar groups of components, such as the branches in parallel resistive circuits. This section also examined the resistance, voltage, and current characteristics of parallel resistive circuits. You will not often encounter parallel resistive circuits as standalone circuits, but will find that understanding their characteristics will be fundamental to working with more complicated circuits that contain them.

12. Kirchhoff's Current Law

12.1 Introduction

Section 9 discussed Gustav Richard Kirchhoff, who formulated two laws of electronics and introduced the law we know today as Kirchhoff's Voltage Law. Kirchhoff's Current Law, sometimes abbreviated KCL, is Kirchhoff's other law and the third fundamental law of electronics. Just as Kirchhoff's Voltage Law expresses a relationship for voltages in a circuit, Kirchhoff's Current Law expresses a relationship between currents in a circuit.

Kirchhoff's Current Law states that the algebraic sum of currents entering and leaving a node (the junction between two or more components) is zero. Simply put, this means that the charge arriving at any point in a circuit must equal the charge leaving it, and at the same rate. An analogy to this would be that of water flowing through a pipe. If 1 gallon of water per minute enters the pipe, 1 gallon of water per minute must leave it.

Kirchhoff's Current Law is itself useful for analyzing simple circuits, but is also fundamental to loop current analysis that you will use to analyze more complex circuits.

Current Polarity

As you learned in Section 9, the terms in an algebraic sum must be signed quantities. For current the sign indicates whether the current is entering or leaving a node. A common convention is to treat currents entering a node as positive quantities, and currents leaving a node as negative quantities. Refer to Figure 12-1.

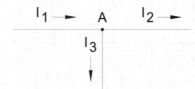

Figure 12-1: Node Currents

Figure 12-1 has three node currents. Kirchhoff's Current Law states that the algebraic sum of these currents equals zero. Using the convention that currents entering a node are positive and that currents leaving a node are negative, you can express this as

$$+I_1 + (-I_2) + (-I_3) = 0$$

The signs you choose for currents entering and leaving a node are not important, provided that

1) the sign of currents entering a node is opposite the sign of currents leaving a node, and

2) you are consistent when using signs.

In this workbook currents entering a node are always positive and currents leaving a node are always negative.

In this section you will:

- Verify the validity of Kirchhoff's Current Law.

- Use Kirchhoff's Current Law to analyze a circuit.

12.2 Pre-Lab

For the circuit of Figure 12-2, use Kirchhoff's Current Law and Ohm's Law to complete the values for the "Calculated" column of Table 12-1.

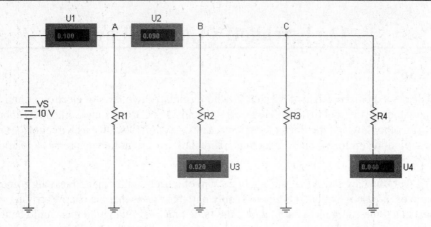

Figure 12-2: Kirchhoff's Current Law Example Circuit

12.3 Design Verification

Open the circuit file Ex12-01. Simulate the circuit and complete the values for the "Measured" column of Table 12-1.

Table 12-1: Calculated and Measured Values for Kirchhoff's Current Law Example Circuit

Circuit Value	Calculated	Measured
VS	10 V	
I_T	0.100 A	
I_{R1}		
$R1$		
I_{AB}	0.090 A	
I_{R2}	0.020 A	
$R2$		
I_{BC}		
I_{R3}		
$R3$		
I_{R4}	0.040 A	
$R4$		

12.4 Application Exercise

Consider the circuit in Figure 12-3. This is a parallel resistive circuit that consists of the two current sources I_{S1} and I_{S2}, the two resistors $R1$ and $R2$, and the six unique node currents I_1 through I_6.

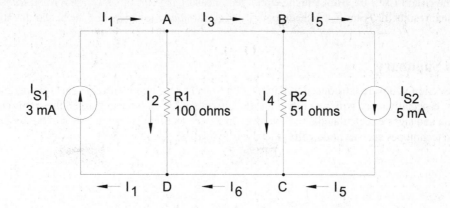

Figure 12-3: Demonstration of Kirchhoff's Current Law

The directions of the node currents at this point are assumptions, although some (such as I_1 and I_5) must follow from the idea of a current source. As with the circuit in Section 9.1, a negative current value will indicate that the actual direction of current flow is opposite to the assumed direction.

Kirchhoff's Current Law for the nodes A, B, C, and D give

Node A: $(+I_1) + (-I_2) + (-I_3) = 0$ Node B: $(+I_3) + (-I_4) + (-I_5) = 0$

Node C: $(+I_4) + (+I_5) + (-I_6) = 0$ Node D: $(+I_2) + (+I_6) + (-I_1) = 0$

From the definition of a current source, $I1 = 3$ mA and $I5 = 5$ mA, so from substitution

Node A: $(+3 \text{ mA}) + (-I_2) + (-I_3) = 0$ Node B: $(+I_3) + (-I_4) + (-5 \text{ mA}) = 0$

Node C: $(+I_4) + (+5 \text{ mA}) + (-I_6) = 0$ Node D: $(+I_2) + (+I_6) + (-3 \text{ mA}) = 0$

Rearranging then gives

Node A: $(+I_2) + (+I_3) = +3$ mA   Node B: $(+I_3) + (-I_4) = +5$ mA

Node C: $(-I_4) + (+I_6) = +5$ mA   Node D: $(+I_2) + (+I_6) = +3$ mA

In addition, since $V_{R1} = V_{R3}$ for parallel circuits,

$$V_{R1} = V_{R2} \rightarrow 100\ \Omega \times I_2 = 51\ \Omega \times I_4$$

So

$$I_2 = 0.51 \times I_4$$

After substituting

Node A: $(+0.51 \times I_4) + (+I_3) = +3$ mA

Node B: $(+I_3) + (-I_4) = +5$ mA

Subtracting the equation for Node A from that of Node B yields

Node B − Node A: $[(+I_3) + (-I_4)] - [(+0.51 \times I_4) + (+I_3)] = (+5 \text{ mA}) - (+3 \text{ mA})$

$$[(+I_3) - (+I_3) + (-I_4) - (+0.51 \times I_4)] = 2 \text{ mA}$$

$$-1.51 \times I_4 = 2 \text{ mA}$$

From this $I_4 = (2 \text{ mA}) / (-1.51) = -1.325$ mA, so that $I_2 = (+0.51 \times I_4) = (+0.51 \times -1.33 \text{ mA}) = -0.676$ mA. Note that the values for both I_2 and I_4 are negative, meaning that the directions of the currents are actually from bottom to top rather than from top to bottom as originally assumed.

Open the circuit Ex12-02. Simulate the circuit and measure the currents through *R1* and *R2* to verify the calculated results for I_2 and I_4. Use the polarity of the ammeter to determine the actual direction for I_2 and I_4.

12.5 Section Summary

In this section you were introduced to Kirchhoff's Current Law and its use in analyzing parallel resistive circuits, especially those with more than one dc current source. You also learned the significance of current direction and signs of calculated circuit values. In a later section these concepts will be important when you use techniques such as node voltage analysis to analyze circuit networks.

13. Loading Effects

13.1 Introduction

In Section 10.4 you discovered that increasing the resistance of a voltage divider could affect the meter reading. This is an example of loading. Loading occurs when a connection to a circuit appreciably affects the characteristics of the circuit. The loading effect in Section 10.4 was an example of meter loading. Another type of loading is source loading.

As its name suggests, the meter measuring some circuit value causes meter loading. All meters have resistance and can affect the resistance of the circuit to which they connect. Refer to Figure 13-1.

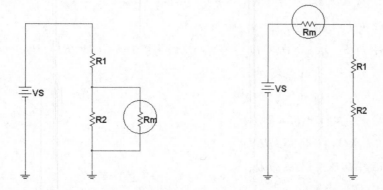

EQUIVALENT CIRCUIT FOR VOLTAGE MEASUREMENTS   EQUIVALENT CIRCUIT FOR CURRENT MEASUREMENTS

Figure 13-1: Equivalent Circuits for Meter Measurements

As the equivalent circuit on the left shows, the voltmeter resistance *Rm* is in parallel with the component whose voltage you wish to measure. This parallel resistance will draw extra current from the circuit and reduce the effective resistance of *R2* and lower the voltage across it. A practical voltmeter has a very high resistance to minimize its effect on the circuit, but if the resistance of the component is close to that of the voltmeter then the meter will noticeably affect the measured voltage.

The circuit on the right shows the situation for current measurements. If *Rm* is low relative to the series resistance of *R1* and *R2* the meter will not greatly affect the circuit resistance and current flowing through it. If the series resistance of the circuit is close to the resistance of the meter, however, *Rm* will change the circuit resistance enough to affect the current.

Source loading occurs when the internal resistance of the source is comparable to the resistance of the circuit to which it connects. Refer to Figure 13-2.

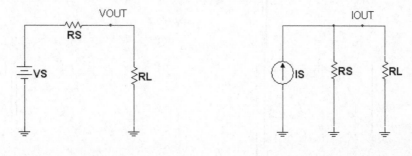

PRACTICAL VOLTAGE SOURCE CIRCUIT   PRACTICAL CURRENT SOURCE CIRCUIT

Figure 13-2: Equivalent Circuits for Practical Sources

The circuit on the left shows the equivalent circuit for a practical voltage source with a load, where *VS* is the ideal voltage source, *RS* is the internal source resistance, *VOUT* is the effective voltage output, and *RL* is the load. From your study of voltage dividers, the effective voltage across *RL* is

$$V_{RL} = VS \times [RL / (RS + RL)]$$

As long as *RL* >> *RS* (the value of *RL* is much larger than that of *RS*), *RL* / (*RS* + *RL*) is close to 1 and the value of $V_{RL} \approx VS$. If *RL* is not large enough to overshadow, or "swamp out," the effect of *RS*, however, then the value of *VOUT* will be noticeably lower than *VS*.

The circuit on the right shows the equivalent circuit for a practical current source with a load, where *IS* is the ideal current source, *RS* is the internal source resistance, *IOUT* is the effective current output, and *RL* is the load. From your work with parallel resistive circuits recall that

$$RS / RL = I_{RL} / I_{RS}$$

From Kirchhoff's Current Law $I_{RS} = IS - I_{RL}$ so

$$RS / RL = (I_{RL}) / (IS - I_{RL}) \rightarrow (IS - I_{RL}) \times (RS / RL) = I_{RL} \rightarrow IS \times (RS / RL) - I_{RL} \times (RS / RL) = I_{RL}$$

From this

$$IS \times (RS / RL) = I_{RL} + [I_{RL} \times (RS / RL)]$$

$$IS \times (RS / RL) = I_{RL} \times [1 + (RS / RL)]$$

$$I_{RL} = IS \times \{(RS / RL) / [1 + (RS / RL)]\}$$

$$I_{RL} = IS \times \{RS / [RL \times (1 + (RS / RL))]\}$$

$$I_{RL} = IS \times [RS / (RL + RS)]$$

If *RL* << *RS* (the value of *RL* is much smaller than *RS*), then the value of *RS* / (*RL* + *RS*) is close to 1 and the value of $I_{RL} \approx IS$. If *RL* becomes large compared to *RS*, however, then *RS* / (*RL* + *RS*) will decrease and the value of I_{RL} will be noticeably lower than the value of *IS*.

In this section you will:

- Study the effects of meter and source loading.

- Experimentally determine the allowable loads for specific levels of loading error.

13.2 Pre-Lab

For the circuit of Figure 13-3, calculate the voltage outputs for the values of *RL* given in Table 13-1 and record them in the "Calculated" column. What is the minimum value of *RL* for which the value of *VOUT* is at least 9 V (90% of *VS*)? How does this value of *RL* compare to *RS* (i.e., what is *RL* / *RS*)?

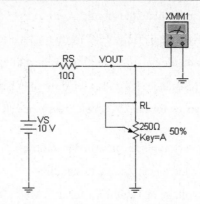

Figure 13-3: Source Loading Example Circuit

13.3 Design Verification

Open the circuit file Ex13-01. Verify your calculated values for each value of *RL* in Table 13-1 and record the values in the "Measured" column of Table 13-1.

Table 13-1: *VOUT* Values for Source Loading Example Circuit

RL Setting	Calculated	Measured	*RL* Setting	Calculated	Measured
0% (0 Ω)			60% (150 Ω)		
10% (25 Ω)			70% (175 Ω)		
20% (50 Ω)			80% (200 Ω)		
30% (75 Ω)			90% (225 Ω)		
40% (100 Ω)			100% (250 Ω)		
50% (125 Ω)					

13.4 Application Exercise

In this exercise you will change the Multisim voltmeter resistance and observe the effects on the measured voltage.

1. Open the circuit file Ex13-02 shown in Figure 13-4.

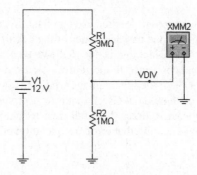

Figure 13-4: Meter Loading Example Circuit

2. Double-click the multimeter to open the enlarged view.

3. Ensure that the meter settings are for the dc voltmeter function.

4. Click the **Run** switch to simulate the circuit.

5. Record the measured voltage in Table 13-2 for the "Meter Resistance" value of 1 GΩ.

6. Click the **Set...** button on the enlarged multimeter view to open the Multimeter Settings window.

7. Click the "Voltmeter resistance (R)" value window and change the value from "1" to "500".

8. Click the drop-down arrow for the "Voltmeter resistance (R)" units and change the units from "GΩ" to "MΩ".

9. Click the **Accept** button to apply the new meter setting.

10. Wait a few seconds for the meter reading to stabilize.

11. Record the measured voltage in Table 13-2 for the "Meter Resistance" value of 500 MΩ.

12. Repeat Steps 6 through 11 for the other values of "Meter Resistance" in Table 13-2.

13. Click the **Run** switch to stop the simulation.

Table 13-2: *VDIV* Measurements for Meter Loading Circuit

Meter Resistance	VDIV	Meter Resistance	VDIV
1 GΩ		5 MΩ	
500 MΩ		2 MΩ	
200 MΩ		1 MΩ	
100 MΩ		500 kΩ	
50 MΩ		200 kΩ	
20 MΩ		100 kΩ	
10 MΩ		50 kΩ	

For what value of meter resistance does *VDIV* drop below 2.700 V (90% of the ideal value of 3.000 V)?

How does this meter resistance compare with *R2* (i.e., what is *Rm / R2*)?

13.5 Section Summary

In this section you examined two types of loading effects. The first type of loading, called meter loading, is due to the meter resistance affecting the characteristics of the circuit that the meter is measuring. The second type of loading, called source loading, is due to the value of internal source resistance becoming appreciable compared to the load connected to the source. You should have determined that the output of a voltage source will be at least 90% of the ideal voltage when the load resistance is at least 10 times the source resistance. Similarly, you should have seen that error from voltmeter loading will be 10% or less when the meter resistance is at least 10 times that of the resistor across which the voltmeter connects. This "10 times" rule is a good one for ensuring that error from loading effects does not exceed 10%.

14. Working with Circuit Equivalents

14.1 Introduction

A common saying is, "A difference that makes no difference is no difference." Circuit equivalents are based upon this idea. A circuit equivalent is a circuit that, within its intended scope, exhibits the same characteristics as another circuit. You have actually used circuit equivalents before, although you may not have realized it. When you calculated the total resistance of a series or parallel circuit to find the total current through the circuit you were actually determining the resistance of a circuit equivalent. Refer to Figure 14-1.

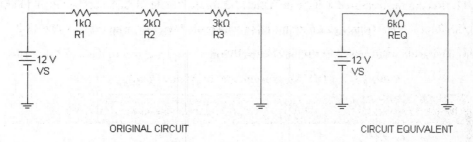

ORIGINAL CIRCUIT   CIRCUIT EQUIVALENT

Figure 14-1: Original Circuit and Circuit Equivalent

The total resistance of the original series resistive circuit is $R_T = 1\text{ k}\Omega + 2\text{ k}\Omega + 3\text{ k}\Omega = 6\text{ k}\Omega$. The total resistance of the circuit equivalent is also 6 kΩ. Although the circuits are not the same, they are equivalent for the intent of finding the total current. Once you know the total current you can find the individual voltage drops across each resistor of the original circuit, something that you cannot do with the circuit equivalent. The power of circuit equivalents is their ability to reduce complex circuits to simpler circuits so that you can apply basic electronic laws and equations to analyze them. When you simplify the circuit, however, you lose detailed circuit information. To analyze the circuit in detail you must work backwards to expand the circuit equivalent back to the original circuit form, applying the information you obtain at each step to obtain more information at the next step.

In this section you will:

* Learn how to develop circuit equivalents from more complex circuits.

* Learn how to expand circuit equivalents to obtain more detailed circuit information.

* Use the Multisim software to verify your calculated values from circuit equivalents.

14.2 Pre-Lab

The first step to developing a circuit equivalent is to find some convenient way to represent circuits. A standard notation uses the "+" operator to represent a series connection and the "||" operator to represent a parallel connection. The "||" operator takes precedence over the "+" operator, just as multiplication and division take precedence over addition and subtraction in mathematical expressions. Just as with mathematical operations, you can use parentheses to override the standard order of operations. Consider the following circuit expressions:

$$RT = (R1 + R2) \parallel R3$$

$$RT = R1 + R2 \parallel R3$$

The first expression represents a circuit in which the series combination of *R1* and *R2* is in parallel with *R3*. The second expression represents a circuit in which *R1* is in series with the parallel combination of *R2* and *R3*. Refer to the circuits in Figure 14-2 that illustrate this.

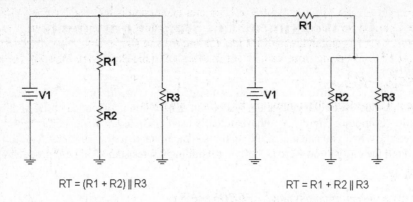

$$RT = (R1 + R2) \parallel R3 \qquad\qquad RT = R1 + R2 \parallel R3$$

Figure 14-2: Comparison of Circuit Expressions

Determine the circuit expressions for the circuits in Figure 14-3 and record them in Table 14-1.

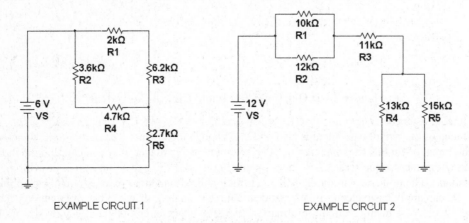

Figure 14-3: Example Circuits

Table 14-1: Circuit Expression for Example Circuits

Circuit	Circuit Expression for *RT*
Example Circuit 1	
Example Circuit 2	

Compare your expressions with those of Table 14-2 and Table 14-3.

14.3 Design Verification

Once you have the expression for a circuit you can simplify the expression and develop a circuit equivalent. The process consists of replacing circuit combinations with their equivalent resistances. The order in which you replace circuit combinations is from left to right using the following order of operations:

1) Parenthetical expressions, working from inner parentheses to outer parentheses

2) Parallel combinations

3) Series combinations

Refer to Example Circuit 1 in Figure 14-3. The circuit expression is $RT = (R1 + R3) \parallel (R2 + R4) + R5$. The order of operations requires you to first evaluate expressions in parentheses to determine the equivalent resistances. The series combinations of $R1$ and $R3$ and $R2$ and $R4$ are both in parentheses so you must evaluate these first. Working from left to right first evaluates the expression $R1 + R3$ to give

$RT = REQ1 \parallel (R2 + R4) + R5$

The next step is to evaluate the expression $R2 + R4$. This gives

$RT = REQ1 \parallel REQ2 + R5$

The "$\parallel$" operator takes precedence over the "+" operator, so the next step is to evaluate $REQ1 \parallel REQ2$.

$RT = REQ3 + R5$

Finally, evaluate the series combination of $REQ3$ and $R5$.

$RT = REQ4$

Refer to Figure 14-4 to see this sequence of operations.

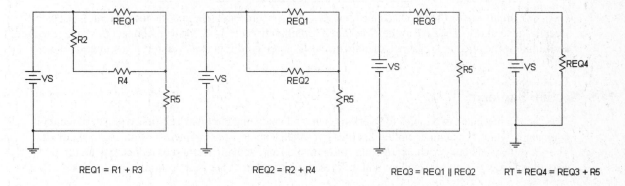

Figure 14-4: Sequence of Circuit Simplification

Once you determine the sequence of operations you can use the equations for total series and parallel resistance to calculate the actual values. Calculate the indicated values in Table 14-2 to determine the equivalent and total resistances for Example Circuit 1 and record them in the "Calculated" column.

Table 14-2: Calculations for Example Circuit 1 Equivalent $RT = (R1 + R3) \parallel (R2 + R4) + R5$

Circuit Value	Calculated	Measured
VS	6 V	
$REQ1 = R1 + R3$		
$REQ2 = R2 + R4$		
$REQ3 = REQ1 \parallel REQ2$		
$RT = REQ4 = REQ3 + R5$		
$IT = VS / RT$		

Open the circuit file Ex14-01. Disconnect the dc voltage source and measure the total circuit resistance. Record the total resistance RT in the "Measured" column in Table 14-2. Then reconnect the voltage source and simulate the circuit. Measure the total current IT and record it in the "Measured" column in Table 14-2.

14.4 Application Exercise

Calculate the indicated values in Table 14-3 to determine the equivalent and total resistances for Example Circuit 2 and record them in the "Calculated" column.

Table 14-3: Calculations for Example Circuit 2 Equivalent $RT = R1 \parallel R2 + R3 + R4 \parallel R5$

Circuit Value	Calculated	Measured
VS	12 V	
$REQ1 = R1 \parallel R2$		
$REQ2 = R4 \parallel R5$		
$REQ3 = REQ1 + R3$		
$RT = REQ4 = REQ3 + REQ2$		
$IT = VS / RT$		

Open the circuit file Ex14-02. Disconnect the dc voltage source and measure the total circuit resistance. Record the total resistance RT in the "Measured" column in Table 14-3. Then reconnect the voltage source and simulate the circuit. Measure the total current IT and record it in the "Measured" column in Table 14-3.

14.5 Section Summary

This section presented two topics for working with and analyzing circuits. The first was that of a circuit equivalent, which is a simpler circuit that preserves the characteristics of interest of the original circuit. The second was that of circuit expressions that provided a systematic method of reducing a circuit to a circuit equivalent. You will explore both of these topics further in the next section.

15. Series-Parallel Resistive Circuits

15.1 Introduction

So far your Multisim circuits have been either series resistive circuits or parallel resistive circuits. In the real world very few circuits are only series or parallel resistive circuits. Most are a combination of the two basic circuit types, or series-parallel resistive circuits. Figure 15-1 shows a simple series-parallel resistive circuit.

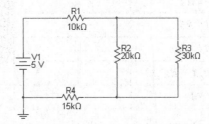

Figure 15-1: Example of Series-Parallel Resistive Circuit

Here *R1* and *R4* are in series with each other, but neither is in series or parallel with either *R2* or *R3*. Similarly, *R2* and *R3* are in parallel with each other, but neither is in series or parallel with either *R1* or *R4*. For this circuit the parallel combination of *R2* and *R3* is in series with *R1* and *R4*. You cannot use the formulas for total series or total parallel resistance alone to find the total circuit resistance, or to determine the internal voltages and currents. Instead, you must first work through the series and parallel combinations to progressively simplify the circuit. The circuit expressions and circuit equivalents you studied in Section 14 are one way to do this. Once you have developed the circuit equivalents you can then work backwards, using the equivalent resistances to determine the voltages and currents for each component.

In this section you will:

- Practice converting series-parallel resistive circuits to equivalent circuits.

- Analyze currents and voltages in series-parallel circuits.

- Compare calculated and measured component voltage and current values to troubleshoot series-parallel circuits.

15.2 Pre-Lab

Refer to the series-parallel resistive circuit in Figure 15-2.

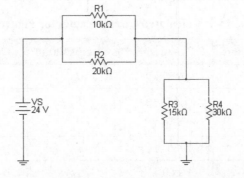

Figure 15-2: Example Series-Parallel Circuit

The expression for this circuit is $RT = R1 \parallel R2 + R3 \parallel R4$. Use the given circuit values to complete Table 15-1. Figure 15-3 shows the sequence of simplifying the circuit.

Table 15-1: Circuit Equivalent Calculations for Example Series-Parallel Circuit

Circuit Value	Calculated
$REQ1 = R1 \parallel R2$	
$REQ2 = R3 \parallel R4$	
$RT = REQ3 = REQ1 + REQ2$	

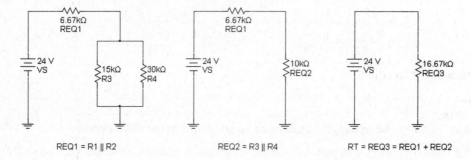

Figure 15-3: Sequence of Generating Circuit Equivalents

Each of the circuit equivalents contains one or more equivalent resistances. Although these resistances do not exist as actual components, you can treat them as such for circuit calculations just as you can calculate the total current from the equivalent resistance RT. You can use this to calculate the voltage and current for each component in the circuit. To do so, start from the last equivalent circuit and work backwards.

In the last circuit equivalent, $RT = REQ = 16.67$ kΩ. Therefore $I_T = VS / RT = 24$ V / 16.67 kΩ = 1.44 mA.

As the previous circuit equivalent shows, $RT = REQ1 + REQ2$, so $I_{REQ1} = I_{REQ2} = I_T$. Therefore $V_{REQ1} = I_{REQ1} \times REQ1 = I_T \times REQ1 = 1.44$ mA × 6.67 kΩ = 9.60 V. Similarly, $V_{REQ2} = I_{REQ2} \times REQ2 = I_T \times REQ2 = 1.44$ mA × 10 kΩ = 14.4 V.

From the previous circuit equivalent, $REQ2 = R3 \parallel R4$ so that $V_{R3} = V_{R4} = V_{REQ2}$. Therefore, $I_{R3} = V_{R3} / R3 = V_{REQ2} / R3 = 14.4$ V / 15 kΩ = 960 µA. Similarly, $I_{R4} = V_{R4} / R4 = V_{REQ2} / R4 = 14.4$ V / 30 kΩ = 480 µA.

And finally, from the original circuit, $REQ1 = R1 \parallel R2$ and $V_{R1} = V_{R2} = V_{REQ1}$. Therefore, $I_{R1} = V_{R1} / R1 = V_{REQ1} / R1 = 9.60$ V / 10 kΩ = 960 µA and $I_{R2} = V_{R2} / R2 = V_{REQ1} / R2 = 9.60$ V / 20 kΩ = 480 µA.

Table 15-2 shows the calculated circuit values for the circuit of Figure 15-2.

Table 15-2: Calculated Circuit Values for Figure 15-2

Circuit Value	Calculated	Circuit Value	Calculated
RT	16.67 kΩ	I_{R2}	480 µA
IT	1.44 mA	V_{R3}	14.4 V
V_{R1}	9.60 V	I_{R3}	960 µA
I_{R1}	960 µA	V_{R4}	14.4 V
V_{R2}	9.60 V	I_{R4}	480 µA

15.3 Design Verification

Open the circuit file Ex15-01. Simulate the circuit and record the circuit values to complete Table 15-3.

Table 15-3: Measured Circuit Values for Figure 15-2

Circuit Value	Measured	Circuit Value	Measured
RT		I_{R2}	
IT		V_{R3}	
V_{R1}		I_{R3}	
I_{R1}		V_{R4}	
V_{R2}		I_{R4}	

15.4 Application Exercise

Figure 15-4 shows a special type of series-parallel circuit called a ladder circuit. A ladder circuit is a circuit that consists of two or more sections, each of which is similar to the others. In the circuit of Figure 15-4 *R1* and *R2* make up the first section, *R3* and *R4* make up the second, and *R5* and *R6* make up the third.

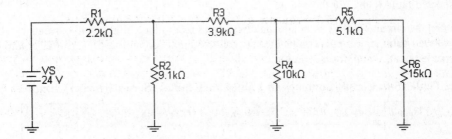

Figure 15-4: Ladder Circuit

The circuit expression for this ladder circuit is

$$RT = R1 + R2 \parallel (R3 + R4 \parallel (R5 + R6))$$

Use the values in Figure 15-4 to complete Table 15-4 with the values for the circuit equivalents.

Table 15-4: Circuit Equivalent Calculations for Figure 15-4

Circuit Value	Calculated
$REQ1 = R5 + R6$	
$REQ2 = R4 \parallel REQ1$	
$REQ3 = R3 + REQ2$	
$REQ4 = R2 \parallel REQ3$	
$RT = REQ5 = R1 + REQ4$	

Use the technique described in the pre-lab section for analyzing the series-parallel circuit in Figure 15-2 to work backwards and calculate the circuit values for the ladder circuit. Record them in the "Calculated" columns of Table 15-5.

Table 15-5: Calculated and Measured Circuit Values for Figure 15-4

Circuit Value	Calculated	Measured	Circuit Value	Calculated	Measured
RT			I_{R3}		
IT			V_{R4}		
V_{R1}			I_{R4}		
I_{R1}			V_{R5}		
V_{R2}			I_{R5}		
I_{R2}			V_{R6}		
V_{R3}			I_{R6}		

Open the circuit file Ex15-02. Simulate the circuit and record the circuit values in the "Measured" columns of Table 15-5.

15.5 Troubleshooting Circuits

Copy the indicated current and voltage values from Table 15-2 into the "No Fault" column of Table 15-6. Then open the circuit Tr15-01, simulate the circuit, and record the voltages and currents for the circuit in the "Tr15-01" column. Compare the values for the "No Fault" column and "Tr15-01" column and indicate the suspected fault, if any, for the circuit.

Next, open the circuit Tr15-02, simulate the circuit, and record the voltages and currents for the circuit in the "Tr15-02" column. Compare the values for the "No Fault" column and "Tr15-02" column and indicate the suspected fault, if any, for the circuit.

Table 15-6: Circuit Comparison Values for Example Series-Parallel Circuit 1

Circuit Value	No Fault	Tr15-01	Tr15-02
I_{R1}			
V_{R1}			
I_{R2}			
V_{R2}			
I_{R3}			
V_{R3}			
I_{R4}			
V_{R4}			
Suspected Fault (if any)			

15.6 Section Summary

This section demonstrated how to use circuit expressions and circuit equivalents that you studied in Section 14 to analyze series-parallel resistive circuits. An equivalent circuit simplifies the form of the circuit and permits you to use the basic circuit equations you have learned thus far, such as Ohm's Law, to calculate voltages and currents in the circuit. Once you have calculated the current and voltage for each resistor in the circuit, you can then compare these circuit values to measured values to isolate and identify possible faults in the circuit.

16. Superposition Theorem

16.1 Introduction

Most circuits have a single power source, but some circuits will have more than one. The Superposition Theorem allows you to analyze circuits that contain more than one power source, provided that they consist of linear bilateral components. A linear bilateral component is one for which

1) current is directly proportional to the applied voltage (doubling the voltage will double the current), and

2) the polarity of the applied voltage does not affect the magnitude of the current (current flows equally well in both directions).

Resistors are linear bilateral components, so the Superposition Theorem applies to resistive circuits.

Put simply, the Superposition Theorem states that the net current through any component is equal to the algebraic sum of the currents due to the individual sources acting independently. This means that you can find the currents in a multi-source circuit by calculating the currents due to each power source separately and then summing the currents.

Note, however, that each source appears as a resistance in the circuit to the other sources. When you calculate the current for each source you must take these other source resistances into account. To do this, you must replace other sources in the circuit with their equivalent resistances when calculating the currents for a source. For voltage sources this resistance is a short (0 Ω), while for current sources this resistance is an open (infinite resistance).

In this section you will:

- Learn how to use the superposition method to analyze multi-source circuits.

- Use the Multisim software to verify your calculated results from the superposition method.

16.2 Pre-Lab

Consider the circuit in Figure 16-1. What are the currents through **R1**, **R2**, and **R3**?

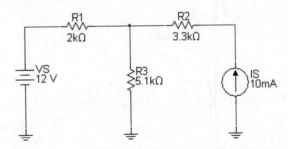

Figure 16-1: Example Superposition Circuit

Step 1: **Determine which current directions through resistors are positive and which are negative.**
As long as you are consistent it doesn't matter which directions have which signs, but for this example currents flowing left to right and top to bottom through a resistor are positive, and those flowing right to left and bottom to top through a resistor are negative.

Step 2: **Select a source and replace all other sources with their internal resistances.** For this step select **VS** as the source, and replace **IS** with its internal resistance. Because **IS** is a current source its internal resistance is infinite, or an open circuit. Figure 16-2 shows the resulting circuit.

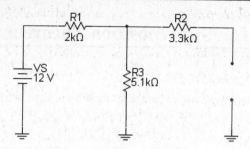

Figure 16-2: Superposition Circuit to Analyze *VS*

Step 3: **Determine the currents due to the selected source.** For this circuit no current from *VS* flows through *R2* because of the open circuit that replaced *IS,* so that *RT* = *R1* + *R3* = 2 kΩ + 5.1 kΩ = 7.1 kΩ. *IT* = *VS* / *RT* = (12 V) / (7.1 kΩ) = 1.69 mA. Current flows left to right through *R1* and top to bottom through *R3*, so $I_{R1}(VS)$ = +1.69 mA and $I_{R3}(VS)$ = +1.69 mA.

Step 4: **Repeat Steps 2 and 3 for all other sources.** This circuit has only one other source, namely *IS*. Select *IS* as the source and replace *VS* with its internal resistance. Because *VS* is a voltage source its internal resistance is 0 Ω, or a short circuit. Figure 16-3 shows the resulting circuit.

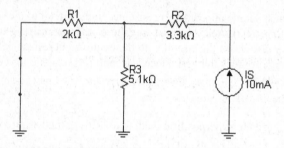

Figure 16-3: Superposition Circuit to Analyze *IS*

The circuit expression for the total resistance for *IS* is *RT* = *R2* + *R1* ‖ *R3*. The circuit equivalent for this circuit, shown in Figure 16-4, is *RT* = *R2* + *REQ* where *REQ* = *R1* ‖ *R3* = 2 kΩ ‖ 5.1 kΩ = 1.44 kΩ.

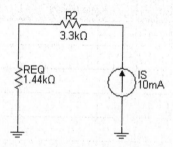

Figure 16-4: Superposition Circuit Equivalent to Analyze *IS*

Because *R2* and *REQ* are in series the same current flows through both. This current is the 10 mA current from *IS*, so that $V_{REQ} = I_{REQ} \times REQ = IS \times REQ$ = 10 mA × 1.44 kΩ = 14.4 V. *REQ* is the parallel combination of *R1* and *R3* so $V_{REQ} = V_{R1} = V_{R3}$. Then $I_{R1} = V_{R1} / R1$ = (14.4 V) / (2 kΩ) = 7.2 mA and $I_{R3} = V_{R3} / R3$ = (14.4 V) / (5.1 kΩ) = 2.82 mA. Current flows right to left through *R1*, right to left through *R2*, and top to bottom through *R3* so $I_{R1}(IS)$ = −7.2 mA, $I_{R2}(IS)$ = −10 mA, and $I_{R3}(IS)$ = +2.82 mA.

Step 5: Sum the currents from all sources. From the calculations for I_{R1}, I_{R2}, and I_{R3} the total currents are

$$I_{R1} = I_{R1}(VS) + I_{R1}(IS) = (+1.69 \text{ mA}) + (-7.2 \text{ mA}) = -5.51 \text{ mA}$$

$$I_{R2} = I_{R2}(VS) + I_{R2}(IS) = (0 \text{ mA}) + (-10 \text{ mA}) = -10 \text{ mA}$$

$$I_{R3} = I_{R3}(VS) + I_{R3}(IS) = (+1.69 \text{ mA}) + (+2.82 \text{ mA}) = +4.51 \text{ mA}$$

The signs of the answers indicate that the current directions are from right to left through **R1** and **R2** and from top to bottom through **R3**, as shown in Figure 16-5.

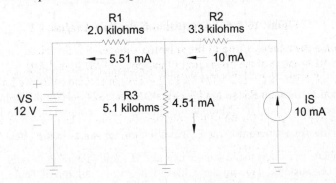

Figure 16-5: Current Directions for Example Superposition Circuit

As a check, do the currents satisfy Kirchhoff's Current Law?

Use Ohm's Law to calculate the voltage across each resistor and record the current and voltage values in the "Calculated" column of Table 16-1.

16.3 Design Verification

Open the circuit file Ex16-01. Run the circuit simulation, measure the current and voltage for each resistor, and record the values in the "Measured" column of Table 16-1. Do the measured values agree with the calculated values?

Table 16-1: Circuit Values for Superposition Example Circuit

Circuit Value	Calculated	Measured
I_{R1}		
V_{R1}		
I_{R2}		
V_{R2}		
I_{R3}		
V_{R3}		

16.4 Application Exercise

Refer to the circuit of Figure 16-6. Assume that currents flowing from left to right and from top to bottom through resistors are positive, and that currents flowing from right to left and from bottom to top through resistors are negative.

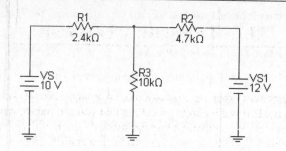

Figure 16-6: Superposition Practice Circuit

Using the superposition method

1) Calculate the resistor currents for *VS1* and record the values in the "VS1" column of Table 16-2.

2) Use the value of each resistor and the calculated value of current through it to calculate the voltage across each resistor and record the values in the "VS1" column of Table 16-2.

3) Calculate the resistor currents for *VS2* and record the values in the "VS2" column of Table 16-2.

4) Use the value of each resistor and the calculated value of current through it to calculate the voltage across each resistor and record the values in the "VS2" column of Table 16-2.

5) Calculate the algebraic sum of the resistor currents and record the values in the "Calculated" column of Table 16-2.

6) Use the value of each resistor and the algebraic sum of current through it to calculate the voltage across each resistor and record the value in the "Calculated" column of Table 16-2.

    Do the voltages in the "VS1" and "VS2" columns sum to the voltages in the "Calculated" column?

Open the circuit file Pr16-01. Measure the voltage and current values for each resistor and record the values in the "Measured" column of Table 16-2.

Table 16-2: Calculated and Measured Values for Superposition Practice Circuit

Circuit Value	*VS1*	*VS2*	Calculated	Measured
I_{R1}				
V_{R1}				
I_{R2}				
V_{R2}				
I_{R3}				
V_{R3}				

Do the voltages in the "Calculated" and "Measured" columns agree?

16.5 Section Summary

This section introduced the Superposition Theorem, which provides a means for calculating the currents and voltages in multi-source resistive circuits. The superposition method analyzes the circuit values for each source independently and then sums the values due to each source to determine the total circuit values.

17. Thevenin's Theorem

17.1 Introduction

The circuit equivalent you will probably use most often to analyze circuits is the Thevenin equivalent. The basis for this circuit equivalent is a theorem that states that you can reduce any two-terminal linear bilateral network to a two-terminal circuit equivalent that consists of a single voltage source in series with a single resistor. The German scientist Hermann von Helmholtz first stated a form of this theorem in a paper in 1853, and it is sometimes referred to as Helmholtz's Theorem or the Helmholtz-Thevenin theorem. Generally, however, most credit the theorem to French telegraph engineer Léon Charles Thévenin who published his own paper in 1883, and refer to the theorem as Thevenin's Theorem. This theorem is actually an extension of the Superposition Theorem, which is why it applies only to linear bilateral circuits.

The great advantage of a Thevenin equivalent is that it allows you to analyze the circuit effects due to one component by reducing the remainder of the circuit to a simple voltage source and series resistor. Refer to Figure 17-1.

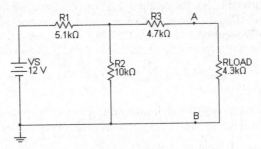

Figure 17-1: Sample Ladder Circuit

You could use the circuit analysis techniques you've learned to analyze the ladder circuit in Figure 17-1 and determine V_{AB} and I_{RLOAD}. But if **RLOAD** changes, you must analyze the entire circuit all over again even if you are not interested in such circuit values as I_{R1} or V_{R3}. Refer now to Figure 17-2.

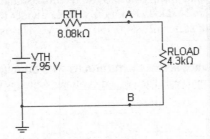

Figure 17-2: Circuit Equivalent

If **RLOAD** for Figure 17-2 changes you can calculate the value of V_{AB} and I_{RLOAD} much more easily than you can for Figure 17-1. The circuit equivalent has consolidated the individual effects of the components to the left of V_{AB} so that you need not consider them if the value of only **RLOAD** changes. In addition, the circuit as a whole is now a simple two-resistor series circuit that is far easier to analyze than the original ladder circuit. The circuit equivalent of **VTH** and **RTH** is the Thevenin equivalent for the ladder circuit of Figure 17-1.

In this section you will:

- Learn how to create a Thevenin equivalent for ("Thevenize") a circuit.

- Use the Multisim software to verify the validity of the Thevenin equivalent.

17.2 Pre-Lab

This example will demonstrate how to create a Thevenin circuit to analyze the effects of **RLOAD** for the circuit shown in Figure 17-3.

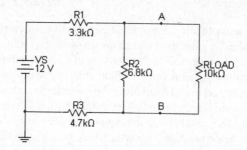

Figure 17-3: Thevenin Equivalent Example Circuit

Step 1: **Identify the terminals for which you wish to create the Thevenin equivalent and remove the component of interest.** For the circuit in Figure 17-3 these are the terminals to which **RLOAD** connects, as **RLOAD** is the component of interest, and are labeled A and B. Figure 17-4 shows the circuit to be Thevenized.

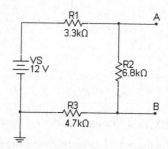

Figure 17-4: Circuit with *RLOAD* Removed

Step 2: **Determine the open load voltage across the terminals.** For this circuit $R_T = R1 + R2 + R3 = 3.3$ kΩ + 6.8 kΩ + 4.7 kΩ = 14.8 kΩ. From the voltage divider formula $V_{AB} = V_{R2} = VS \times (R2 / R_T) =$ (12 V) × (6.8 kΩ / 14.8 kΩ) = 12 V × 0.460 = 5.51 V. This is the open load voltage across A and B so $V_{TH} = 5.51$ V.

Step 3: **Determine the resistance across the terminals with all sources replaced by their internal resistances.** The internal resistance of a dc voltage source is 0 Ω, so replace **VS** with a wire, as shown in Figure 17-5.

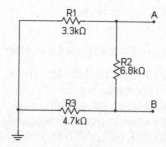

Figure 17-5: Circuit with *VS* Replaced by Short and *RLOAD* Removed

The expression for the resistance seen between terminals A and B is $R_{AB} = R2 \parallel (R1 + R3) = 6.8$ k$\Omega \parallel (3.3$ k$\Omega + 4.7$ k$\Omega) = 6.8$ k$\Omega \parallel 8.0$ k$\Omega = 3.68$ kΩ. This is the resistance between A and B with the voltage source replaced by its internal resistance, so $R_{TH} = 3.68$ kΩ.

From the above calculations, the Thevenin equivalent for the circuit of Figure 17-4 is a 5.51 Vdc voltage source in series with a 3.68 kΩ, as shown in Figure 17-6.

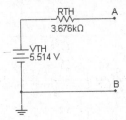

Figure 17-6: Thevenin Equivalent for Figure 17-4

Calculate the voltages and currents indicated in Table 17-1 for the original circuit in Figure 17-3 and record your answers in the "Calculated" column.

Table 17-1: *RLOAD* Calculations for Figure 17-3

Circuit Values	Calculated
$REQ = R2 \parallel RLOAD$	
$R_T = R1 + REQ + R3$	
$V_{RLOAD} = V_{REQ} = VS \times (REQ / R_T)$	
$I_{RLOAD} = V_{RLOAD} / RLOAD$	

Calculate the voltages and currents indicated in Table 17-2 for the Thevenized circuit in Figure 17-6 with RLOAD attached and record your answers in the "Calculated" column.

Table 17-2: Calculations for *RLOAD* in Figure 17-6

Circuit Value	Calculated
$R_T = RTH + RLOAD$	
$V_{RLOAD} = V_{TH} \times (RLOAD / R_T)$	
$I_{RLOAD} = V_{RLOAD} / RLOAD$	

How do the answers for V_{RLOAD} and I_{RLOAD} in Table 17-1 and Table 17-2 compare?

17.3 Data Verification

Open the circuit file Ex17-01. Simulate the circuit for each of the values of **RLOAD** listed in Table 17-3 and record the values of V_{RLOAD} and I_{RLOAD} in the "Original Circuit" column.

Next, open the circuit file Ex17-02. Simulate the circuit for each of the values of **RLOAD** listed in Table 17-3 and record the values of V_{RLOAD} and I_{RLOAD} in the "Thevenin Equivalent" column.

Table 17-3: Comparison of Measured *RLOAD* Voltages and Currents

RLOAD	Original Circuit		Thevenin Equivalent	
	V_{RLOAD}	I_{RLOAD}	V_{RLOAD}	I_{RLOAD}
100 Ω				
200 Ω				
500 Ω				
1 kΩ				
2 kΩ				
5 kΩ				
10 kΩ				
20 kΩ				
50 kΩ				
100 kΩ				

How do the measured currents and voltages compare for the original and Thevenin equivalent circuits?

17.4 Application Exercise

Although the best approach to finding the Thevenin equivalent is to calculate the values, you can use the Multisim tools as a shortcut to find the Thevenin voltage and resistance of a circuit by directly measuring the open load voltage and equivalent resistance.

1. Open the circuit file Ex17-03 shown in Figure 17-7.

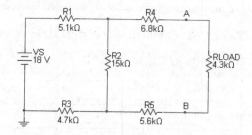

Figure 17-7: Application Exercise Circuit

2. Detach ***RLOAD*** from the circuit.

3. Connect the multimeter across terminals A and B.

4. Click **Run** switch to start the circuit simulation.

5. Measure the dc voltage across terminals A and B and record this value as V_{TH} in Table 17-4.

6. Click the **Run** switch to stop the simulation.

7. Delete the dc voltage source ***VS***.

8. Use a wire to connect the left terminal of ***R1*** to ground to simulate the internal resistance of ***VS***.

9. Change the multimeter mode to the ohmmeter setting.

10. Click the **Run** switch to start the circuit simulation.

11. Measure the resistance across terminals A and B and record this value as R_{TH} in Table 17-4.

12. Click the **Run** switch to stop the simulation.

13. Close the circuit file without saving your changes.

Table 17-4: Application Exercise Circuit Thevenin Equivalent Values

V_{TH}	R_{TH}

Open the circuit file Ex17-03 again. Simulate the circuit for each of the values of **RLOAD** listed in Table 17-5 and record the values of V_{RLOAD} and I_{RLOAD} in the "Original Circuit" column.

Next, create a Thevenin equivalent circuit using your values from Table 17-4 and connect a load resistor across the circuit. Simulate the circuit for each of the values of **RLOAD** listed in Table 17-5 and record the values of V_{RLOAD} and I_{RLOAD} in the "Thevenin Equivalent" column.

Table 17-5: Original and Thevenized Application Circuit Measurements

RLOAD	Original Circuit		Thevenin Equivalent	
	V_{RLOAD}	I_{RLOAD}	V_{RLOAD}	I_{RLOAD}
100 Ω				
200 Ω				
500 Ω				
1 kΩ				
2 kΩ				
5 kΩ				
10 kΩ				
20 kΩ				
50 kΩ				
100 kΩ				
200 kΩ				
500 kΩ				

How do the measured currents and voltages compare for the original circuit and your calculated Thevenin equivalent circuit?

17.5 Section Summary

This section introduced the Thevenin equivalent, which is a circuit equivalent that reduces a circuit to a voltage source (called the Thevenin voltage) and series resistor (called the Thevenin resistance). The Thevenin equivalent allows you to more easily determine the effects of changing the value of a circuit component on its voltage and current without recalculating the voltages and currents for the rest of the circuit. In future sections you will see the value of the Thevenin equivalent in analyzing such parameters as power transfer and circuit time constants.

18. Norton's Theorem

18.1 Introduction

In 1926 Hans Ferdinand Mayer, a Hause-Siemens researcher, and Edward Lawry Norton, a Bell Labs engineer, independently introduced an extension of Thevenin's Theorem. Although Norton reported his findings only in an internal publication while Mayer published his work, the theorem on which they both worked is known today as Norton's Theorem. The Norton equivalent is based on this theorem, which states that you can reduce any two-terminal linear bilateral network to a two-terminal circuit equivalent that consists of a single current source in parallel with a single resistor. Refer to Figure 18-1.

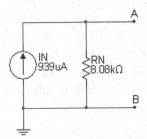

Figure 18-1: Example Norton Equivalent Circuit

As with a Thevenin equivalent, the advantage of the Norton equivalent is that it allows you to analyze the circuit effects due to one component by reducing the remainder of the circuit to a simple current source and parallel resistor. Norton equivalents are useful for analyzing circuits in which current is the parameter of interest, such as transistor circuits.

In this section you will:

- Learn how to create a Norton equivalent for ("Nortonize") a circuit.

- Use the Multisim software to verify the validity of the Norton equivalent.

- Develop a Norton equivalent from a Thevenin equivalent and vice versa.

18.2 Pre-Lab

This example will demonstrate how to create a Norton circuit to analyze the effects of **RLOAD** for the circuit shown in Figure 18-2, which is the same example circuit that Section 17.2 used.

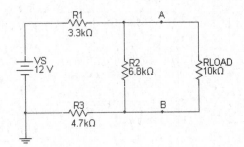

Figure 18-2: Norton Equivalent Example Circuit

Step 1: **Identify the terminals for which you wish to create the Norton equivalent and remove the component of interest.** For the circuit in Figure 18-2 these are the terminals to which **RLOAD** connects, as **RLOAD** is the component of interest, and are labeled A and B. Figure 18-3 shows the circuit to be Nortonized.

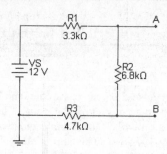

Figure 18-3: Circuit with *RLOAD* Removed

Step 2: **Determine the short circuit current between the terminals.** For this circuit, imagine a wire between terminals A and B in Figure 18-3. Since R2 is bypassed by the short circuit, $R_T = R1 + R3 = 3.3\ k\Omega + 4.7\ k\Omega = 8.0\ k\Omega$. Then $I_{AB} = 12\ V / 8.0\ k\Omega = 1.5\ mA$. Since this is the short circuit current between terminals A and B, $I_N = 1.5\ mA$.

Step 3: **Determine the resistance across the terminals with all sources replaced by their internal resistances.** The internal resistance of a dc voltage source is 0 Ω, so replace *VS* with a wire, as shown in Figure 18-4.

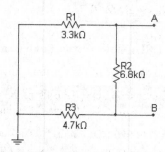

Figure 18-4: Circuit with *VS* Replaced by Short and *RLOAD* Removed

The expression for the resistance seen between terminals A and B is $R_{AB} = R2 \parallel (R1 + R3) = 6.8\ k\Omega \parallel (3.3\ k\Omega + 4.7\ k\Omega) = 6.8\ k\Omega \parallel 8.0\ k\Omega = 3.676\ k\Omega$. This is the resistance between A and B with the voltage source replaced by its internal resistance, so $R_N = 3.676\ k\Omega$. Note that the Norton resistance is the same as the Thevenin resistance for the circuit.

The Norton equivalent for the circuit of Figure 18-3 is a 1.5 mA dc current source in parallel with a 3.68 kΩ resistor, as shown in Figure 18-5.

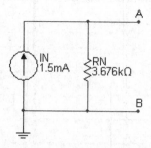

Figure 18-5: Norton Equivalent for Figure 18-3

Calculate the voltages and currents indicated in Table 18-1 for the original circuit in Figure 18-2 and record your answers in the "Calculated" column.

Table 18-1: *RLOAD* **Calculations for Figure 18-2**

Circuit Values	Calculated
$REQ = R2 \parallel RLOAD$	
$R_T = R1 + REQ + R3$	
$V_{RLOAD} = V_{REQ} = VS \times (REQ / R_T)$	
$I_{RLOAD} = V_{RLOAD} / RLOAD$	

Calculate the voltages and currents indicated in Table 18-2 for the Nortonized circuit in Figure 18-5 with *RLOAD* attached and record your answers in the "Calculated" column.

Table 18-2: Calculations for *RLOAD* **in Figure 18-5**

Circuit Value	Calculated
$R_T = R_N \parallel RLOAD$	
$V_{RLOAD} = V_T = I_N \times R_T$	
$I_{RLOAD} = V_{RLOAD} / RLOAD$	

How do the answers for V_{RLOAD} and I_{RLOAD} in Table 18-1 and Table 18-2 compare?

18.3 Data Verification

Open the circuit file Ex18-01. Simulate the circuit for each of the values of *RLOAD* listed in Table 18-3 and record the values of V_{RLOAD} and I_{RLOAD} in the "Original Circuit" column.

Next, open the circuit file Ex18-02. Simulate the circuit for each of the values of *RLOAD* listed in Table 18-3 and record the values of V_{RLOAD} and I_{RLOAD} in the "Norton Equivalent" column.

Table 18-3: Comparison of Measured *RLOAD* **Voltages and Currents**

RLOAD	Original Circuit		Norton Equivalent	
	V_{RLOAD}	I_{RLOAD}	V_{RLOAD}	I_{RLOAD}
100 Ω				
200 Ω				
500 Ω				
1 kΩ				
2 kΩ				
5 kΩ				
10 kΩ				
20 kΩ				
50 kΩ				
100 kΩ				

How do the measured currents and voltages compare for the original and Norton equivalent circuits?

18.4 Application Exercise

Although the best approach to finding the Norton equivalent is to calculate the values, you can use the Multisim tools as a shortcut to find the Norton current and resistance of a circuit by directly measuring the short circuit current and equivalent resistance.

1. Open the circuit file Ex18-03 shown in Figure 18-6.

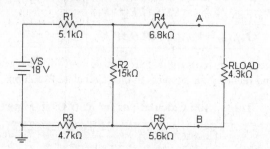

Figure 18-6: Application Exercise Circuit

2. Detach **RLOAD** from the circuit.

3. Connect the multimeter between terminals A and B.

4. Change the multimeter mode to the dc ammeter setting.

5. Click the **Run** switch to start the circuit simulation.

6. Measure the dc current between terminals A and B and record this value as I_N in Table 18-4.

7. Click the **Run** switch to stop the simulation.

8. Delete the dc voltage source **VS**.

9. Use a wire to connect the left terminal of **R1** to ground to simulate the internal resistance of **VS**.

10. Change the multimeter mode to the ohmmeter setting.

11. Click the **Run** switch to start the circuit simulation.

12. Measure the resistance across terminals A and B and record this value as R_N in Table 18-4.

13. Click the **Run** switch to stop the simulation.

14. Close the circuit file without saving your changes.

Table 18-4: Application Exercise Circuit Norton Equivalent Values

I_N	R_N

Open the circuit file Ex18-03 again. Simulate the circuit for each of the values of **RLOAD** listed in Table 18-5 and record the values of V_{RLOAD} and I_{RLOAD} in the "Original Circuit" column.

Next, create a Norton equivalent circuit using your values from Table 18-5 and connect a load resistor across the circuit. Simulate the circuit for each of the values of **RLOAD** listed in Table 18-5 and record the values of V_{RLOAD} and I_{RLOAD} in the "Norton Equivalent" column.

Table 18-5: Original and Nortonized Application Circuit Measurements

RLOAD	Original Circuit		Norton Equivalent	
	V_{RLOAD}	I_{RLOAD}	V_{RLOAD}	I_{RLOAD}
100 Ω				
200 Ω				
500 Ω				
1 kΩ				
2 kΩ				
5 kΩ				
10 kΩ				
20 kΩ				
50 kΩ				
100 kΩ				
200 kΩ				
500 kΩ				

How do the measured currents and voltages compare for the original circuit and your calculated Norton equivalent circuit?

One interesting aspect of Thevenin and Norton equivalents is that you can easily derive one from the other.

Refer to Figure 18-5 for the procedure to convert a Norton equivalent to a Thevenin equivalent. The Thevenin voltage is the open load voltage across terminals A and B, so for the Norton equivalent $V_{AB} = V_{TH} = IN \times RN = 1.5$ mA × 3.676 kΩ = 5.514 V. The Thevenin resistance, on the other hand, is the resistance between terminals A and B with the source replaced by its internal resistance. Since the internal resistance of a current source is infinite, $R_{TH} = R_{AB} = RN = 3.68$ kΩ. The procedure for calculating the Norton resistance is identical to that for calculating the Thevenin resistance, so this should not surprise you.

To convert a Thevenin equivalent to a Norton equivalent, consider from the above that in general $R_N = R_{TH}$. Further, $V_{TH} = I_N \times R_N$, so that $I_N = V_{TH} / R_N$. Since $R_N = R_{TH}$, however, then $I_N = V_{TH} / R_{TH}$.

Use the current and resistance values for the Norton equivalent in Figure 18-5 to calculate the voltage and resistance values for the Thevenin equivalent. Do the values agree with those in Figure 17-6?

18.5 Section Summary

This section introduced the Norton equivalent, which is a circuit equivalent that reduces a circuit to a current source (called the Norton current) and parallel resistor (called the Norton resistance). The Norton equivalent allows you to more easily determine the effects of varying the value of a circuit component on its voltage and current without recalculating the voltages and currents for the rest of the circuit. Norton equivalents are primarily used for circuits in which currents are the parameter of interest. You can use the following equations to convert a Norton equivalent to a Thevenin equivalent and vice versa:

$R_{TH} = R_N$

$V_{TH} = I_N \times R_N$

$I_N = V_{TH} / R_{TH}$

19. Maximum Power Transfer

19.1 Introduction

Every electronic circuit dissipates power. A primary goal of electronic design is to ensure that a circuit delivers as much power as possible to the intended load. This is a simple task if the load is the only element in the circuit to dissipate power, but practical circuits contain other components that also dissipate power. Refer to the circuit shown in Figure 19-1.

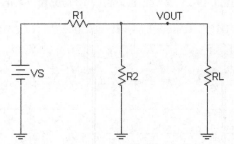

Figure 19-1: Power Transfer Example Circuit

Resistors **R1** and **R2** form a voltage divider to develop the voltage **VOUT**, which the circuit applies to the load resistor **RL**. The power that the circuit delivers to **RL** is $P_{RL} = VOUT \times I_{RL}$, and the circuit ideally delivers as much power as possible to **RL**. For what value of **RL** will this occur?

Consider the extreme case when the value of **RL** is infinite. **RL** will not load the voltage divider so **VOUT** is as large as it can possibly be, but I_{RL} is 0 A. Consequently $P_{RL} = VOUT(\text{max}) \times 0 \text{ A} = 0 \text{ W}$. If the value of **RL** decreases then I_{RL} will start to increase, but **RL** will begin to load the voltage divider and cause **VOUT** to decrease. Consequently P_{RL} will initially increase, but at some point the loading effect of **RL** will dominate and P_{RL} will start to decrease.

Now consider the other extreme for which **RL** = 0 Ω. Now I_{RL} is as large as it can possibly be but **VOUT** = 0 V, so that $P_{RL} = 0 \text{ V} \times I_{RL}(\text{max}) = 0 \text{ W}$. If the value of **RL** increases then the loading effect will decrease so that **VOUT** will increase, but the larger values of **RL** will cause the value of I_{RL} to decrease. Consequently P_{RL} will initially increase, but at some point the resistance of **RL** will dominate and P_{RL} will start to decrease.

From the above analysis it appears that P_{RL} should be low for a very low value of **RL**, increase to some maximum as **RL** increases, and then decrease as **RL** continues to increase. The **maximum power transfer theorem** states that a resistive circuit transfers maximum power to the load when the load resistance is equal to the effective source resistance. If the load resistance is less than this value then loading effects dominate, and if the load resistance is greater than this value then the load resistance dominates.

In this section you will:

- Use the Multisim software to experimentally verify the maximum power transfer theorem.

- Use Thevenin's Theorem to determine the ideal load resistance for maximum power transfer.

19.2 Pre-Lab

For the circuit shown in Figure 19-2, calculate the Table 19-1 column values for each of the indicated values of load resistance. Then plot P_{RL}, the power dissipated by the load resister, against **RL**, the value of load resistance.

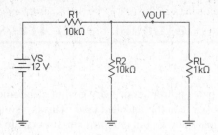

Figure 19-2: Power Transfer Evaluation Circuit

Table 19-1: Calculate Load Resistance Power Dissipation

RL	$R_{EQ} = R2 \parallel RL$	$R_T = R1 + R_{EQ}$	$VOUT = VS \times (R_{EQ} / R_T)$	$I_{RL} = VOUT/RL$	$P_{RL} = VOUT \times I_{RL}$
1 kΩ					
2 kΩ					
3 kΩ					
4 kΩ					
5 kΩ					
6 kΩ					
7 kΩ					
8 kΩ					
9 kΩ					
10 kΩ					

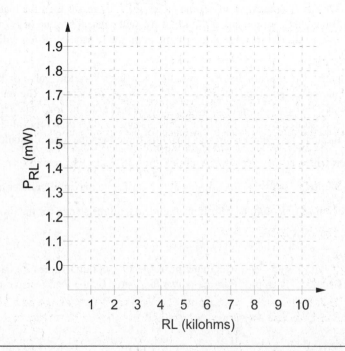

For approximately what value of load resistance is the power in *RL* maximum?

19.3 Design Verification

Open and simulate the circuit file Ex19-01 shown in Figure 19-3. Set the value of potentiometer *RL* to each of the values shown in Table 19-2 and record the values measured by the wattmeter.

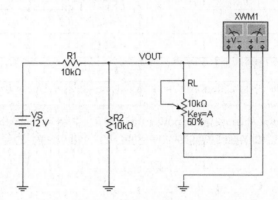

Figure 19-3: Verification Circuit for Power Transfer Evaluation Circuit

Table 19-2: Measured Power for Verification Circuit

RL	Measured Power	*RL*	Measured Power
1 kΩ		6 kΩ	
2 kΩ		7 kΩ	
3 kΩ		8 kΩ	
4 kΩ		9 kΩ	
5 kΩ		10 kΩ	

Do the measured values agree with your calculated values in Table 19-1?

19.4 Application Exercise

The maximum power transfer theorem states that a resistive circuit delivers maximum power to the load when the load resistance is equal to the effective source resistance. Your calculations and measurements for the verification circuit should have shown that power in *RL* is maximum when the value of *RL* is about 5 kΩ. This means that the effective source resistance for the circuit is about 5 kΩ. Even though neither *R1* or *R2* is 5 kΩ, the effective source resistance for the circuit is 5 kΩ because the resistance of the Thevenin (and Norton) equivalent is 5 kΩ. For the circuit of Figure 19-2, the Thevenin voltage and resistance are

$V_{TH} = VS \times [R2 / (R1 + R2)]$

$= 12 \text{ V} \times [10 \text{ k}\Omega / (10 \text{ k}\Omega + 10 \text{ k}\Omega)]$

$= 12 \text{ V} \times (10 \text{ k}\Omega / 20 \text{ k}\Omega) = 6 \text{ V}$

$R_{TH} = R1 \parallel R2$

$= 10 \text{ k}\Omega \parallel 10 \text{ k}\Omega = 5 \text{ k}\Omega$

Open the circuit file Pr19-01, which contains the Thevenin equivalent for the power transfer evaluation circuit of Figure 19-2. Simulate the circuit and record the wattmeter values for each of the values of *RL* indicated in Table 19-3.

Do the values agree with those of Table 19-1 and Table 19-2?

Table 19-3: Measured Power for Thevenin Equivalent

RL	Measured Power	RL	Measured Power
1 kΩ		6 kΩ	
2 kΩ		7 kΩ	
3 kΩ		8 kΩ	
4 kΩ		9 kΩ	
5 kΩ		10 kΩ	

You can always determine the maximum power transfer to a load by Thevenizing the circuit that connects to the load. The circuit equivalent then becomes that shown in Figure 19-4.

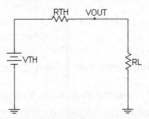

Figure 19-4: Thevenized Power Transfer Circuit

The circuit will deliver maximum power to the load when $RL = RTH$. When this is true, it follows that the power in the load is equal to the power dissipated by the rest of the circuit, which is represented by the Thevenin resistance. It is also true that $VOUT = VTH / 2$. Since VTH is the open load voltage, you can experimentally determine when the circuit delivers maximum power to the load by adjusting the load resistance until the value of $VOUT$ is half the value of the open load voltage.

Refer now to the circuit in Figure 19-5.

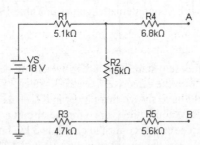

Figure 19-5: Maximum Power Transfer Practice Circuit

Calculate the Thevenin voltage and resistance for the circuit and record them in Table 19-4.

Table 19-4: Calculated Values for Practice Circuit Thevenin Equivalent

V_{TH}	R_{TH}

Next, you will experimentally determine the resistance value for maximum power transfer. In this procedure you will use successive approximation to determine the resistor value for which the output voltage is half the open load voltage.

1. Open the circuit file Pr19-02.

2. Click the **Run** switch to start the circuit simulation.

3. Use the multimeter to measure the open load voltage V_{AB} across terminals A and B. This voltage should equal your calculated Thevenin voltage in Table 19-4.

4. Click the **Run** switch to stop the circuit simulation.

5. Connect a virtual resistor as the load between terminals A and B.

6. Click the **Run** switch to start the circuit simulation.

7. Measure the voltage across the load resistor.

8. Click the **Run** switch to stop the circuit simulation.

9. If the voltage measurement in Step 7 is greater than half the open load voltage you measured in Step 3, reduce the resistor value. If the voltage measurement in Step 7 is less than half the open load voltage you measured in Step 7, increase the resistor value. One method is to use a binary adjustment method. Start by increasing the resistor value by 1 kΩ until the measured voltage is greater than half the open load voltage. Then halve the change in resistor value to 500 Ω and reduce the resistor value until the measured voltage is less than half the open load voltage. Halve the change in resistor value again to 250 Ω and increase the resistance value until the measured voltage is greater than half the open load voltage. Continue to reduce the change in resistor value and zero in on the final resistance value by alternately increasing and decreasing the resistor value each time you must change the direction in which you are moving the resistor value.

10. Repeat Steps 6 through 9 until the load voltage is approximately equal to half the open load voltage you measured in Step 3.

How does the final value of the load resistor compare to your calculated value of Thevenin resistance R_{TH} in Table 19-4?

19.5 Section Summary

In this section you investigated some of the aspects of power transfer in resistive circuits. A circuit will deliver maximum power to the load when the load resistance is equal to the effective source resistance. The effective source resistance is the same as the Thevenin resistance. When a circuit delivers maximum power to the load, the load voltage is equal to half the open load voltage. You can experimentally determine the resistance for maximum power transfer by adjusting the resistance until the voltage across the load is equal to half the open load voltage.

20. Branch Current Analysis

20.1 Introduction

The circuit analysis techniques in this workbook so far have been mostly sequential and incremental in nature. In other words, they have worked through the circuit one equation at a time, with the solution to each equation providing information to solve the next equation. Each equation, based on a specific law or theorem, solved for a single variable. Mathematicians refer to these types of equations as "one equation in one unknown." This approach is suitable for analyzing simple circuits, but can be quite tedious or even inadequate for solving more complicated circuits.

Fortunately, more advanced techniques exist to analyze circuits. Although the implementations of each technique differ, they all share the following characteristics:

1) They merge basic electronic laws in their equations.

2) They work with systems of circuit equations rather than individual equations.

This section will discuss the branch current analysis method. A branch is a circuit path that connects any two nodes, and a node is a connection of two or more components. The branch current analysis method identifies the currents in each branch of the circuit and then uses Ohm's Law and Kirchhoff's Laws to develop these currents into a system of "*n* equations in *n* unknowns" that you can solve using linear algebra techniques.

In this section you will:

- Learn to identify branch currents in a circuit.

- Learn to develop a system of equations from the branch currents in a circuit.

- Learn to represent the system of equations in matrix-vector form.

- Use the Multisim software to verify your solutions from branch current analysis.

20.2 Pre-Lab

Refer to the branch current example circuit in Figure 20-1.

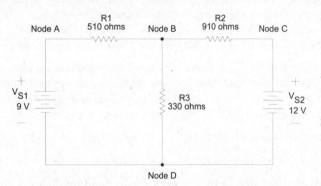

Figure 20-1: Branch Current Example Circuit

Step 1: Identify the branch currents. As the figure shows, there are four circuit nodes, labeled A through D. From the definition of branch current, there are five branch currents between the following nodes in the circuit:

- Node A and Node B (call it I_1)

- Node B and Node C (call it I_2)

- Node B and Node D (call it I_3)

- Node A and Node D (call it I_4)

- Node C and Node D (call it I_5)

V_{S1} and R_1 are in series between Node B and Node C so that the same current must flow through both of them. Therefore I_4 is the same as I_1 (i.e., is redundant) and can be ignored. Similarly, V_{S2} and R_2 are in series between Node B and Node D, so that I_5 is the same as I_2 and also can be ignored. Therefore there are three non-redundant branch currents: I_1, I_2, and I_3.

Step 2: **Select a reference direction for each branch current.** As for the analysis techniques you studied in earlier sections the directions you choose are not critical. If the actual direction for a branch current is opposite that of the reference direction then the calculated current value will be negative to indicate this. For this example, assume that I_1 flows from Node A to Node B, I_2 flows from Node C to Node B, and I_3 flows from Node B to Node D.

Step 3: **Assign polarities for the resistor voltages based on the reference current directions.** Based on the reference currents from Step 2 the current and voltage polarities should be those in Figure 20-2.

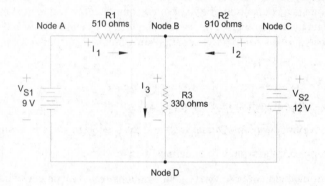

Figure 20-2: Branch Current Example Circuit with Assigned Currents and Voltages

Step 4: **Apply Kirchhoff's Voltage Law to each non-redundant circuit loop.** A non-redundant circuit loop is one that contains at least one voltage that is not already contained in any other circuit loop. For the example circuit, there are two non-redundant circuit loops. The first contains V_{S1}, V_{R1}, and V_{R3}. The second contains V_{S2}, V_{R2}, and V_{R3}. The loop containing V_{S1}, V_{R1}, V_{R2}, and V_{S2} is a redundant circuit loop, as each voltage already appears in another loop.

When applying Kirchhoff's Voltage Law you can choose to sum the voltages in a clockwise or counter-clockwise direction. You can also choose the sign of each voltage to be the sign of the voltage as you encounter the component or as you leave it. For this example, sum the voltages in a clockwise direction and use the sign of the voltage as you encounter each component.

For Loop 1, starting at Node D: $\qquad -V_{S1} + (R1 \times I_1) + (R3 \times I_3) = 0$

For Loop 2, starting at Node D: $\qquad -(R3 \times I_3) + -(R2 \times I_2) + V_{S2} = 0$

Step 5: **Apply Kirchhoff's Current Law to each non-redundant node.** A non-redundant node is one that contains at least one current that is not already accounted for at any other node. For the circuit in Figure 20-2 Node B is the only non-redundant node, as each current present at Node A, Node C, and Node D is also present at Node B.

When applying Kirchhoff's Current Law you can choose currents entering the node to be positive and those exiting the node to be negative, or vice versa. For this example, choose currents entering the node to be positive and those leaving the node to be negative.

For Node B: $(+I_1) + (+I_2) + (-I_3) = 0$

Step 6: Solve the system of equations from Steps 4 and 5. The subject of linear algebra is beyond the scope of this workbook, which will cover only the mechanics of setting up and solving the equations. For more information refer to your electronics textbook or consult your instructor.

From Steps 4 and 5, the system of equations is

$$-V_{S1} + (R1 \times I_1) + (R3 \times I_3) = 0$$

$$-(R3 \times I_3) + -(R2 \times I_2) + V_{S2} = 0$$

$$(+I_1) + (+I_2) + (-I_3) = 0$$

Rearrange the equations to isolate the current terms on the left side.

$$(R1 \times I_1) + (R3 \times I_3) = V_{S1}$$

$$-(R2 \times I_2) + -(R3 \times I_3) = -V_{S2}$$

$$(+I_1) + (+I_2) + (-I_3) = 0$$

Expand the equations so that each equation contains I_1, I_2, and I_3 terms.

$$(R1 \times I_1) + (0 \times I_2) + (R3 \times I_3) = V_{S1}$$

$$(0 \times I_1) + -(R2 \times I_2) + -(R3 \times I_3) = -V_{S2}$$

$$(+I_1) + (+I_2) + (-I_3) = 0$$

Substitute in the circuit resistance and source values.

$$(510\ \Omega \times I1) + (0\ \Omega \times I2) + (330\ \Omega \times I3) = 9\text{ V}$$

$$(0\ \Omega \times I1) + (-910\ \Omega \times I2) + (-330\ \Omega \times I3) = -12\text{ V}$$

$$(1\ \Omega \times I1) + (1\ \Omega \times I2) + (-1\ \Omega \times I3) = 0\text{ V}$$

Express this system in matrix form, as shown below.

$$\begin{vmatrix} 510\ \Omega & 0\ \Omega & 330\ \Omega \\ 0\ \Omega & -910\ \Omega & -330\ \Omega \\ 1\ \Omega & 1\ \Omega & -1\ \Omega \end{vmatrix} \cdot \begin{vmatrix} I1 \\ I2 \\ I3 \end{vmatrix} = \begin{vmatrix} 9\text{ V} \\ -12\text{ V} \\ 0\text{ V} \end{vmatrix}$$

At this point you can solve the system with standard linear algebra techniques, the matrix functions on most scientific calculators, mathematics application software, or most spreadsheet programs. The folder containing the circuit files for this section, as well as the folders for Sections 21 and 22, includes a README.TXT and two .XLS spreadsheet files with which you can solve matrices for 2x2, 3x3, and 4x4 systems. For this system the solution is $I_1 = 7.72$ mA, $I_2 = 7.62$ mA, and $I_3 = 15.3$ mA.

Use the calculated branch current values to complete Table 20-1.

Table 20-1: Calculated Values for Branch Current Example Circuit

Circuit Value	Calculated	Ohm's Law	Calculated	Kirchhoff's Laws	Calculated
I1		$V_{R1} = I1 \times R1$		$I1 + I2 - I3$	
I2		$V_{R2} = I2 \times R2$		$-VS1 + V_{R1} + V_{R3}$	
I3		$V_{R3} = I3 \times R3$		$VS2 - V_{R2} - V_{R3}$	

Do the calculated values satisfy Kirchhoff's Voltage and Current Laws?

20.3 Design Verification

Open the circuit file Ex20-01. Simulate the circuit, measure the voltage and current for each resistor, and record your measured values in Table 20-2. Do the calculated and measured values agree?

Table 20-2: Measured Values for Branch Current Example Circuit

Circuit Value	Measured	Circuit Value	Measured
$I_{R1} = I_1$		V_{R1}	
$I_{R2} = I2$		V_{R2}	
$I_{R3} = I_{R3}$		V_{R3}	

20.4 Application Exercise

Refer to the circuit in Figure 20-3 and use the branch current analysis method to determine the resistor current and voltage values.

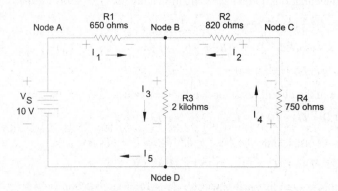

Figure 20-3: Branch Current Application Exercise Circuit

Step 1: Identify branch currents. You should be able to verify that I_4 and I_5 are redundant.

Step 2: Select a reference direction for each branch current. For this circuit use the directions shown in Figure 20-3.

Step 3: Assign polarities for the resistor voltages based on the reference current directions. Verify that the polarities shown in Figure 20-3 are consistent with the reference currents.

Step 4: Apply Kirchhoff's Voltage Law to each non-redundant circuit loop. Write the loop equations for the following two voltage loops. Loop 1 contains V_S, R_1, and R_3. Loop 2 contains R_2, R_3, and R_4. For this example, start at Node D, sum the voltages in a clockwise direction, and use the sign of the voltage as you encounter them for each component.

Loop 1:

Loop 2:

Step 5: Apply Kirchhoff's Current Law to each non-redundant node. You should be able to verify that Node D is redundant.

Current equation:

Step 6: Solve the system of equations from Steps 4 and 5. Verify that you can express your system of equations in the following matrix form:

$$
\begin{vmatrix} 650\ \Omega & 0\ \Omega & 2000\ \Omega \\ 0\ \Omega & -1570\ \Omega & -2000\ \Omega \\ 1\ \Omega & 1\ \Omega & -1\ \Omega \end{vmatrix} \cdot \begin{vmatrix} I1 \\ I2 \\ I3 \end{vmatrix} = \begin{vmatrix} 10\ \text{V} \\ 0\ \text{V} \\ 0\ \text{V} \end{vmatrix}
$$

Record your calculated currents and voltages for **R1**, **R2**, **R3**, and **R4** in the "Calculated" column in Table 20-3.

Table 20-3: Calculated and Measured Values for Branch Current Application Exercise Circuit

Circuit Value	Calculated	Measured	Circuit Value	Calculated	Measured
I_{R1}			V_{R1}		
I_{R2}			V_{R2}		
I_{R3}			V_{R3}		
I_{R4}			V_{R4}		

Open the circuit file Pr20-01. Measure the currents and voltages for **R1**, **R2**, **R3**, and **R4** and record them in the "Measured" column in Table 20-3.

Do the measured and calculated values in Table 20-3 agree?

20.5 Troubleshooting Exercises

Open each of the circuits Tr20-01 through Tr20-03 and complete Table 20-4 with your measured values. Compare these values with your values in Table 20-2 and indicate whether or not a fault exists in each circuit. If a fault exists, indicate the probable nature of the fault.

Table 20-4: Measured Circuit Values for Troubleshooting Circuits

Circuit Value	Tr20-01	Tr20-02	Tr20-03
I_{R1}			
V_{R1}			
I_{R2}			
V_{R2}			
I_{R3}			
V_{R3}			
Suspected Fault (if any)			

20.6 Section Summary

This section discussed the branch current analysis method, which combined basic electronic laws to create a system of equations from the currents through branches in a circuit. Once you reduce the circuit to a system of linear equations, you can use linear algebra methods, calculator functions, or software to solve for the branch currents. With these currents you can then use Ohm's Law to calculate voltages across components in the circuit.

21. Loop Current Analysis

21.1 Introduction

The loop current analysis method is another method for using a system of equations to analyze circuits. This method is similar to the branch current analysis method, but uses reference currents around a closed circuit loop rather than currents between nodes to define the system equations. Each loop current must be non-redundant, which means that the current must flow through at least one component not included in any other loop current. Refer to the circuit in Figure 21-1.

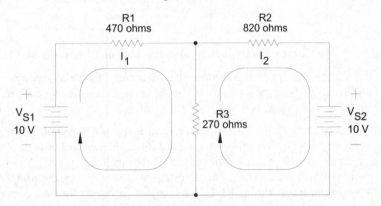

Figure 21-1: Loop Current Example Circuit

The circuit in Figure 21-1 contains two non-redundant loop currents. I_1 passes through V_{S1}, **R1**, and **R3** and I_2 passes through **R2**, **R3**, and V_{S2}. The loop that passes through V_{S1}, **R1**, **R2**, and V_{S2} is redundant, as I_1 or I_2 already passes through every component through which it passes. As with branch current analysis, the direction of the reference current is unimportant, as the sign of the calculated current will indicate whether or not the direction of the reference current is correct.

In this section you will:

- Learn to set up loop currents in a circuit.

- Learn to develop a system of equations from the loop currents in a circuit.

- Learn to represent the system of equations in matrix-vector form.

- Use the Multisim software to verify your solutions from loop current analysis.

21.2 Pre-Lab

This example will use the circuit of Figure 21-1 to demonstrate the loop current analysis method.

Step 1: Identify the loop currents. The direction for these loop currents can be clockwise or counterclockwise, but must be non-redundant. This example will use clockwise current directions for both loops, as shown in Figure 21-1.

Step 2: Assign polarities for the resistor voltages based on the reference current directions. Based on the clockwise directions chosen for Figure 21-1 the voltage polarities are as shown in Figure 21-2.

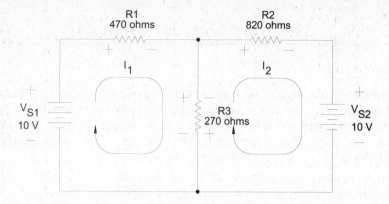

Figure 21-2: Loop Current Example Circuit with Assigned Currents and Voltages

Note that **R3** has two assigned polarities. This is because the voltage polarity due to I_1 differs from that due to I_2. You must be sure to use the correct polarities for each reference current if more than one current passes through a component.

Step 3: **Apply Kirchhoff's Voltage Law to each circuit loop.** When applying Kirchhoff's Voltage Law you can choose to sum the voltages in a clockwise or counterclockwise direction, although the simplest choice is to use the same direction as that of the loop current. You can also choose the sign of each voltage to be the sign of the voltage as you encounter the component or as you leave it. For this example, use the sign of the voltage you encounter for each component.

If more than one loop current passes through a component, you must include the voltage drop due to each loop current. For each loop current use the appropriate voltage polarity that the loop current develops.

For Loop Current I_1: $-V_{S1} + (R1 \times I_1) + (R3 \times I_1) + (-R3 \times I_2) = 0$

For Loop Current I_2: $(R3 \times I_2) + (-R3 \times I_1) + (R2 \times I_2) + V_{S2} = 0$

Step 4: **Solve the system of equations from Step 3.** The system of equations is

$$-V_{S1} + (R1 \times I_1) + (R3 \times I_1) + (-R3 \times I_2) = 0$$

$$(R3 \times I_2) + (-R3 \times I_1) + (R2 \times I_2) + V_{S2} = 0$$

Rearrange the terms to isolate the current terms on the left side.

$$(R1 \times I_1) + (R3 \times I_1) + (-R3 \times I_2) = V_{S1}$$

$$(R3 \times I_2) + (-R3 \times I_1) + (R2 \times I_2) = -V_{S2}$$

Combine the current terms in each equation.

$$[(R1 + R3) \times I_1] + (-R3 \times I_2) = V_{S1}$$

$$(-R3 \times I_1) + [(R2 + R3) \times I_2] = -V_{S2}$$

Substitute in the resistance and source voltage values.

$$[(470\ \Omega + 270\ \Omega) \times I_1] + (-270\ \Omega \times I_2) = 10\ \text{V}$$

$$(-270\ \Omega \times I_1) + [(820\ \Omega + 270\ \Omega) \times I_2] = -10\ \text{V}$$

Combine resistance values in each equation.

$$(740\ \Omega \times I_1) + (-270\ \Omega \times I_2) = 10\ \text{V}$$

$$(-270\ \Omega \times I_1) + (1090\ \Omega \times I_2) = -10\ \text{V}$$

Express this system in matrix form, as shown below.

$$\begin{vmatrix} 740\ \Omega & -270\ \Omega \\ -270\ \Omega & 1090\ \Omega \end{vmatrix} \cdot \begin{vmatrix} I1 \\ I2 \end{vmatrix} = \begin{vmatrix} 10\ \text{V} \\ -10\ \text{V} \end{vmatrix}$$

At this point you can solve the system with standard linear algebra techniques, the matrix functions on most scientific calculators, mathematics application software, or most spreadsheet programs. For this system the solution is $I_1 = 11.2$ mA and $I_2 = -6.41$ mA.

Note that the sign of I_2 is negative and indicates that the direction of the reference current is actually counterclockwise. Note also that while this method used only two equations, rather than the three equations that the branch current method required, it did not directly provide the value of current through $R3$. This is a fundamental difference between the branch current and loop current analysis methods. The branch current method will directly provide the current through each component in the circuit. The loop current method requires you to calculate the current for components through which two or more loop currents pass.

If we define the direction of I_1 as positive then $I_{R3} = I_1 - I_2 = 11.2$ mA $- (-6.41$ mA$) = 17.6$ mA. Because the answer is positive the direction of I_3 is in the same direction as I_1, because the direction of I_1 by definition is positive. If we define the direction of I_2 as positive then $I_{R3} = I_2 - I_1 = -6.41$ mA $- 11.2$ mA $= -17.6$ mA. Because the answer is negative the direction of I_{R3} is opposite to the direction of I_2, because the direction of I_2 by definition is positive.

Use the calculated branch current values to complete Table 21-1.

Table 21-1: Calculated Values for Loop Current Example Circuit

Circuit Value	Calculated	Ohm's Law	Calculated	Kirchhoff's Laws	Calculated
I_1		$V_{R1} = I_1 \times R1$		$I1 + I2 - I_{R3}$	
I_2		$V_{R2} = I_2 \times R2$		$-VS1 + V_{R1} + V_{R3}$	
I_{R3}		$V_{R3} = I_{R3} \times R3$		$VS2 - V_{R2} - V_{R3}$	

Do the calculated values satisfy Kirchhoff's Voltage and Current Laws?

21.3 Design Verification

Open the circuit file Ex21-01. Simulate the circuit, measure the voltage and current for each resistor, and record your measured values in Table 21-2. Do the calculated and measured values agree?

Table 21-2: Measured Values for Loop Current Example Circuit

Circuit Value	Measured	Circuit Value	Measured
$I_{R1} = I_1$		V_{R1}	
$I_{R2} = I_2$		V_{R2}	
I_{R3}		V_{R3}	

21.4 Application Exercise

Refer to the circuit in Figure 21-3 and use the loop current analysis method to determine the resistor current and voltage values. This circuit is the same circuit from Section 20.4 so that you can verify that the branch current and loop current analysis methods return consistent results.

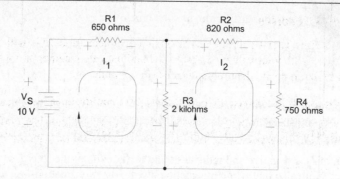

Figure 21-3: Loop Current Application Exercise Circuit

Step 1: **Identify loop currents.** Use the loop currents I_1 and I_2 shown in Figure 21-3.

Step 2: **Assign polarities for the resistor voltages based on the reference current directions.** Verify that the polarities shown in Figure 21-3 are consistent with the reference currents.

Step 3: **Apply Kirchhoff's Voltage Law to each circuit loop.** Write the loop equations for the circuit loops for I_1 and I_2. For this example sum the voltages in the directions of the loop currents, and use the sign of the voltage as you encounter them for each component.

Loop 1:

Loop 2:

Step 4: **Solve the system of equations from Step 3.** Verify that you can express your system of equations in the following matrix form:

$$\begin{vmatrix} 2650\ \Omega & -2000\ \Omega \\ -2000\ \Omega & 3570\ \Omega \end{vmatrix} \cdot \begin{vmatrix} I1 \\ I2 \end{vmatrix} = \begin{vmatrix} 10\ \text{V} \\ 0\ \text{V} \end{vmatrix}$$

Record your calculated currents and voltages for *R1*, *R2*, *R3*, and *R4* in the "Calculated" column in Table 21-3. Note that you must calculate the value of I_{R3} from the values of I_1 and I_2.

Table 21-3: Calculated and Measured Values for Loop Current Application Exercise Circuit

Circuit Value	Calculated	Measured	Circuit Value	Calculated	Measured
I_{R1}			V_{R1}		
I_{R2}			V_{R2}		
I_{R3}			V_{R3}		
I_{R4}			V_{R4}		

Open the circuit file Pr21-01. Measure the currents and voltages for *R1*, *R2*, *R3*, and *R4* and record them in the "Measured" column in Table 21-3.

Do the calculated and measured values in Table 21-3 agree?

Do the loop current and branch current results agree?

21.5 Troubleshooting Exercises

Open each of the circuits Tr21-01 through Tr21-03 and complete Table 21-4 with your measured values. Compare these values with your values in Table 21-2 and indicate whether or not a fault exists in each circuit. If a fault exists, indicate the probable nature of the fault.

Table 21-4: Measured Circuit Values for Troubleshooting Circuits

Circuit Value	Tr21-01	Tr21-02	Tr21-03
I_{R1}			
V_{R1}			
I_{R2}			
V_{R2}			
I_{R3}			
V_{R3}			
Suspected Fault (if any)			

21.6 Section Summary

This section discussed the loop current analysis method, which combined basic electronic laws to create a system of equations from the currents around closed circuit loops. Once you reduce the circuit to a system of linear equations, you can use linear algebra methods, calculator functions, or software to solve for the branch currents. With these currents you can then use Ohm's Law to calculate voltages across components in the circuit. Unlike the branch current analysis method, the loop current method does not directly provide currents through each component in a circuit. You must use the reference currents to determine both the magnitudes and directions of currents for components through which two or more loop currents pass.

22. Node Voltage Analysis

22.1 Introduction

The node voltage analysis method is a variation of the branch current analysis method. Branch current analysis identifies and solves for the currents flowing between the nodes in a circuit. Node voltage analysis identifies and solves for the node voltages between which the branch currents flow. Refer to the circuit in Figure 22-1.

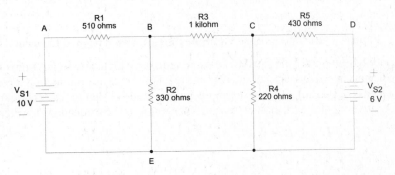

Figure 22-1: Node Voltage Example Circuit

The node voltage example circuit has five unique nodes, identified as A through E. Ohm's Law requires that the branch currents between these nodes equal the difference between the voltages at the nodes divided by the resistance between the nodes. For example, the branch current through *R3* must equal the voltage difference between Nodes B and E divided by 330 Ω. With the node voltage analysis method you assign branch currents, express each one in terms of node voltages and the resistance between them, and use Kirchhoff's Current Law to develop a system of equations for the circuit. You then solve the system of equations for the node voltages.

In this section you will:

- Learn to assign branch currents and node voltages to a circuit.

- Learn to develop a system of equations from the branch currents and node voltages for the circuit.

- Learn to set up the system of equations in matrix-vector form.

- Use the Multisim software to verify your solutions from node voltage analysis.

22.2 Pre-Lab

This example will use the circuit of Figure 22-1 to demonstrate the loop current analysis method.

Step 1: Identify the circuit nodes. The circuit in Figure 22-1 has five circuit nodes, identified by the letters A through E.

Step 2: Select one node as a reference. This node serves the same function as circuit ground, so that all voltage values will be relative to this node. For the circuit of Figure 22-1 an obvious choice is Node E.

Step 3: Assign currents at all nodes for which you do not know the voltages. The voltage at Node A relative to Node E is 10 V because of V_{S1}. Similarly, the voltage at Node B relative to Node E is 6 V because of V_{S2}. Therefore, only the voltages at Nodes B and C are unknown. The directions you choose for the currents at these nodes are unimportant, as the signs of the calculated currents will indicate whether the direction is correct (positive) or incorrect (negative). For this example choose currents directions to be left to right and top to bottom, as shown in Figure 22-2.

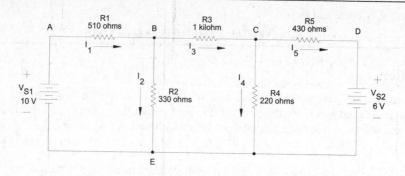

Figure 22-2: Node Voltage Example Circuit with Assigned Currents

Step 4: **Apply Kirchhoff's Current Law to each node for which you do not know the voltage.** The signs you choose for currents entering and leaving a node are not important provided that you use the signs consistently. For this example assign a positive sign for each current entering a node and a negative sign for each current leaving a node. The current equations for Figure 22-2 are

$$(+I_1) + (-I_2) + (-I_3) = 0$$

$$(+I_3) + (-I_4) + (-I_5) = 0$$

Step 5: **Apply Ohm's Law to express each current in terms of the node voltages and circuit resistances.** The current equations are then

$$[(V_A - V_B) / R1] + [-(V_B - V_E) / R2] + [-(V_B - V_C) / R3] = 0$$

$$[(V_B - V_C) / R3] + [-(V_C - V_E) / R4] + [-(V_C - V_D) / R5] = 0$$

Step 6: **Expand and combine terms for each equation.** Expanding terms gives

$$(V_A / R1) + (-V_B / R1) + (-V_B / R2) + (-V_B / R3) + (V_E / R2) + (V_C / R3) = 0$$

$$(V_B / R3) + (-V_C / R3) + (-V_C / R4) + (-V_C / R5) + (V_E / R4) + (V_D / R5) = 0$$

Combining terms gives

$$(V_A / R1) + (V_B) \times -(1/R1 + 1/R2 + 1/R3) + (V_E / R2) + (V_C / R3) = 0$$

$$(V_B / R3) + (V_C) \times -(1/R3 + 1/R4 + 1/R5) + (V_E / R4) + (V_D / R5) = 0$$

Step 7: **Substitute known voltages and resistance values in the equation and simplify.** Substitute voltage and resistance values from Figure 21-1.

$$(10 \text{ V}) / (510 \, \Omega) + (V_B) [-(1/510 \, \Omega + 1/330 \, \Omega + 1/1 \text{ k}\Omega)]$$

$$+ (0 \text{ V}) / (510 \, \Omega) + (V_C) / (1 \text{ k}\Omega) = 0$$

$$(V_B) / (1 \text{ k}\Omega) + (V_C) [-(1/1 \text{ k}\Omega + 1/220 \, \Omega + 1/430 \, \Omega)]$$

$$+ (0 \text{ V}) / (220 \, \Omega) + (6 \text{ V}) / (430 \, \Omega) = 0$$

Remove the 0 terms.

$$(10 \text{ V}) / (510 \, \Omega) + (V_B) [-(1/510 \, \Omega + 1/330 \, \Omega + 1/1 \text{ k}\Omega)] + (V_C) / (1 \text{ k}\Omega) = 0$$

$$(V_B) / (1 \text{ k}\Omega) + (V_C) [-(1/1 \text{ k}\Omega + 1/220 \, \Omega + 1/430 \, \Omega)] + (6 \text{ V}) / (430 \, \Omega) = 0$$

Isolate the V_B and V_C terms on the left side.

$$(V_B) [-(1/510 \, \Omega + 1/330 \, \Omega + 1/1 \text{ k}\Omega)] + (V_C) (1/1 \text{ k}\Omega) = -(10 \text{ V}) / (510 \, \Omega)$$

$$(V_B) (1/1 \text{ k}\Omega) + (V_C) [-(1/1 \text{ k}\Omega + 1/220 \, \Omega + 1/430 \, \Omega] = -(6 \text{ V}) / (430 \, \Omega)$$

Simplify numeric terms.

$$(V_B)(-5.991 \text{ mS}) + (V_C)(1.000 \text{ mS}) = -19.61 \text{ mA}$$

$$(V_B)(1.000 \text{ mS}) + (V_C)(-7.871 \text{ mS}) = -13.95 \text{ mA}$$

Express this system in matrix form, as shown below.

$$\begin{vmatrix} -5.991 \text{ mS} & 1.000 \text{ mS} \\ 1.000 \text{ mS} & -7.871 \text{ mS} \end{vmatrix} \cdot \begin{vmatrix} V_B \\ V_C \end{vmatrix} = \begin{vmatrix} -19.61 \text{ mA} \\ -13.95 \text{ mA} \end{vmatrix}$$

At this point you can solve the system with standard linear algebra techniques, the matrix functions on most scientific calculators, mathematics application software, or most spreadsheet programs. For this system the solution is $V_B = 3.65$ V and $V_C = 2.24$ V.

Use these values of V_A, V_B, V_C, V_D, and V_E to complete Table 22-1.

Table 22-1: Calculated Circuit Values for Node Voltage Example Circuit

Circuit Value	Calculated	Circuit Value	Calculated
V_A	10 V	$I_1 = (V_A - V_B) / 510 \ \Omega$	
V_B	3.65 V	$I_2 = (V_B - V_E) / 330 \ \Omega$	
V_C	2.24 V	$I_3 = (V_B - V_C) / 1 \ \text{k}\Omega$	
V_D	6 V	$I_4 = (V_C - V_E) / 220 \ \Omega$	
V_E	0 V	$I_5 = (V_C - V_D) / 430 \ \Omega$	

Which calculated current directions (if any) are opposite to the directions of the reference currents in Figure 22-2?

22.3 Design Verification

Open the circuit file Ex22-01. Simulate the circuit, measure the voltage and current for each resistor, and record your measured values in Table 22-2.

Table 22-2: Measured Values for Node Voltage Example Circuit

Circuit Value	Measured	Circuit Value	Measured
V_A		I_1	
V_B		I_2	
V_C		I_3	
V_D		I_4	
V_E	0 V	I_5	

Do the calculated values in Table 22-1 and measured values in Table 22-2 agree?

22.4 Application Exercise

Refer to the circuit in Figure 22-3 and use the node voltage analysis method to determine the resistor current and voltage values. This circuit is the same circuit from Sections 20.4 and 21.4 so that you can verify that the branch current, loop current, and node voltage methods return consistent results.

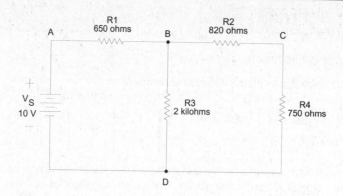

Figure 22-3: Node Voltage Application Exercise Circuit

Step 1: **Identify the circuit nodes.** The circuit in Figure 22-3 has four circuit nodes, identified by the letters A through D.

Step 2: **Select one node as a reference.** For the circuit of Figure 22-3 use Node D.

Step 3: **Assign currents at all nodes for which you do not know the voltages.** For the circuit of Figure 22-3 you should be able to identify two current nodes and four node currents. For this exercise choose current directions to be left to right and top to bottom, as for the example circuit of Figure 22-2.

Step 4: **Apply Kirchhoff's Current Law to each node for which you do not know the voltage.** For this exercise assign a positive sign for each current entering a node and a negative sign for each current leaving a node.

Step 5: **Apply Ohm's Law to express each current in terms of the node voltages and circuit resistances.**

Equation 1:

Equation 2:

Step 6: **Expand and combine terms for each equation.**

Step 7: **Substitute known voltages and resistance values in the equation and simplify.** Verify that you can express your system of equations in the following matrix form:

$$\begin{vmatrix} -3.258 \text{ mS} & 1.220 \text{ mS} \\ 1.220 \text{ mS} & -2.553 \text{ mS} \end{vmatrix} \cdot \begin{vmatrix} V_B \\ V_C \end{vmatrix} = \begin{vmatrix} -15.38 \text{ mA} \\ 0 \text{ mA} \end{vmatrix}$$

Solve for V_B and V_C. Then record your calculated values for V_A, V_B, V_C, and V_D and your calculated voltages and currents for **R1**, **R2**, **R3**, and **R4** in the "Calculated" column in Table 22-3.

Table 22-3: Calculated and Measured Values for Node Voltage Application Exercise Circuit

Circuit Value	Calculated	Measured	Circuit Value	Calculated	Measured
V_A			$I_{R3} = (V_B - V_D) / 2 \text{ k}\Omega$		
V_B			$I_{R4} = (V_C - V_D) / 750 \ \Omega$		
V_C			$V_{R1} = I_{R1} \times 650 \ \Omega$		
V_D			$V_{R2} = I_{R2} \times 820 \ \Omega$		

Circuit Value	Calculated	Measured	Circuit Value	Calculated	Measured
$I_{R1} = (V_A - V_B) / 650 \ \Omega$			$V_{R3} = I_{R3} \times 2 \ k\Omega$		
$I_{R2} = (V_B - V_C) / 820 \ \Omega$			$V_{R4} = I_{R4} \times 750 \ \Omega$		

Open the circuit file Pr22-01. Measure the currents and voltages for *R1*, *R2*, *R3*, and *R4* and record them in the "Measured" column in Table 22-3.

Do the calculated and measured values in Table 22-3 agree?

Do the node voltage, loop current, and branch current results agree?

22.5 Troubleshooting Exercises

Open each of the circuits Tr22-01 through Tr22-03 and complete Table 22-4 with your measured values. Compare these values with your values in Table 22-2 and indicate whether or not a fault exists in each circuit. If a fault exists, identify the probable nature of the fault.

Table 22-4: Measured Circuit Values for Troubleshooting Circuits

Circuit Value	Tr22-01	Tr22-02	Tr22-03
V_A			
V_B			
V_C			
V_D			
V_E			
$I_1 = I_{R1}$			
$I_2 = I_{R2}$			
$I_3 = I_{R3}$			
$I_4 = I_{R4}$			
$I_5 = I_{R5}$			
Suspected Fault (if any)			

22.6 Section Summary

This section discussed the node voltage analysis method, which combined basic electronic laws to create a system of equations from the node voltages present within the circuit. Once you reduce the circuit to a system of linear equations, you can use linear algebra methods, calculator functions, or software to solve for the node voltages. With these voltages you can then use Ohm's Law to calculate currents through components in the circuit. The node voltage method has the advantage of generally having fewer voltages for which to solve than the branch current method and directly providing voltages in the circuit.

23. Bridge Circuits

23.1 Introduction

A special class of series-parallel circuit that you will encounter is the bridge circuit. Figure 23-1 shows two different views of a resistive bridge circuit. The conventional representation on the left shows the way in which most drawings depict a bridge circuit. The orthogonal representation on the right shows the same circuit with the resistors and wiring oriented vertically and horizontally (i.e., orthogonally). The orthogonal representation is the one that you will see in the Multisim program, as the software requires you to use vertical and horizontal orientation for components and wiring.

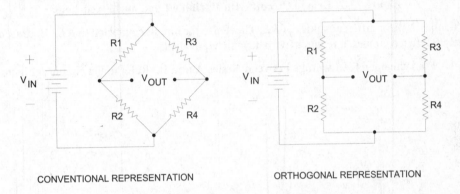

CONVENTIONAL REPRESENTATION   ORTHOGONAL REPRESENTATION

Figure 23-1: Bridge Circuit Views

As the orthogonal representation shows, a resistive bridge circuit consists of two parallel circuits, each of which consists of two series resistances, or "legs". What distinguishes a bridge circuit is that the input voltage is across the parallel circuits, while the output voltage is taken between them. In this way the output "bridges" the two parallel circuits.

Bridge circuits are used extensively to make extremely precise measurements. One common bridge is the Wheatstone bridge, which measures resistance. Similar bridges can measure the inductance and capacitance. The sensitivity of bridges allows them to register the very small resistance changes of strain gages and similar transducers.

In this section you will:

- Learn methods of analyzing resistive bridge circuits.

- Analyze the operation of the Wheatstone bridge.

- Use the Multisim program to verify your analysis of bridge circuits.

23.2 Pre-Lab

Two common problems that relate to bridge circuits are

1) Finding the value of V_{OUT} when the values of the legs are fixed and the value of a load resistor across V_{OUT} changes.

2) Finding the value of current through a fixed resistor when the value of one of the legs changes.

23.2.1 Finding V_{OUT} for a Variable Load Resistor

Refer to the circuit in Figure 23-2, which shows a variable load **RL** connected between nodes A and B. For this circuit $V_{OUT} = V_{RL} = |V_{AB}|$.

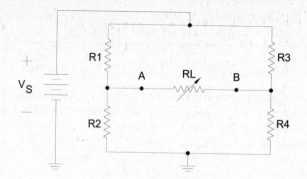

Figure 23-2: Bridge Circuit with Variable Load and Fixed Legs

To calculate V_{RL} for different values of **RL**, Thevenize the circuit connected to **RL**. Because **RL** does not directly reference ground, it requires two Thevenin equivalents.

Step 1: Find the open load voltage between Nodes A and B. Refer to the circuit in Figure 23-3.

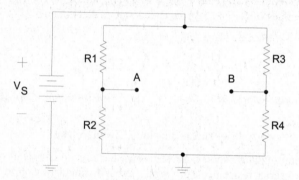

Figure 23-3: Calculating the Open Load Voltages

From voltage divider formula V_{THA}, the Thevenin voltage for Node A is $V_S \times [R2 / (R1 + R2)]$. Similarly, V_{THB}, the Thevenin voltage for Node B is $V_S \times [R4 / (R2 + R4)]$.

Step 2: Find the equivalent resistances for Nodes A and B with *VS* shorted. Refer to the circuit in Figure 23-4.

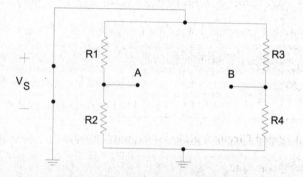

Figure 23-4: Calculating the Equivalent Resistances

The resistance between Node A and ground with V_S shorted is **R1** in parallel with **R2**, so R_{THA}, the Thevenin resistance for Node A, is **R1 ∥ R2**. Similarly, R_{THB}, the Thevenin resistance for Node B, is **R3 ∥ R4**.

From Steps 1 and 2 the Thevenin equivalent of the bridge circuit in Figure 23-2 is the circuit shown in Figure 23-5.

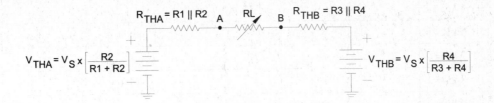

Figure 23-5: Thevenin Equivalent for Figure 23-2

The voltage across R_{THA}, RL, and R_{THB} is $V_{THA} - V_{THB}$, so by Ohm's Law

$$I_{RL} = (V_{THA} - V_{THB}) / (R_{THA} + RL + R_{THB}) \text{ and } V_{RL} = RL \times [(V_{THA} - V_{THB}) / (R_{THA} + RL + R_{THB})].$$

For the circuit of Figure 23-6, calculate V_{THA}, R_{THA}, V_{THB}, and R_{THB} and record the values in Table 23-1. Then, calculate the values of I_{RL} and V_{RL} for each value of RL in Table 23-2.

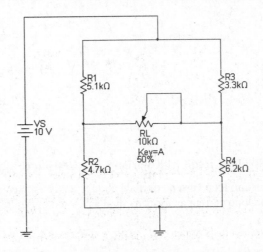

Figure 23-6: Example Bridge Circuit with Variable Load

Table 23-1: Calculated Thevenin Values for Example Bridge Circuit with Variable Load

Circuit Value	Calculated	Circuit Value	Calculated
$V_{THA} = VS \times [R2/(R1 + R2)]$		$V_{THB} = VS \times [R4/(R3 + R4)]$	
$R_{THA} = R1 \parallel R2$		$R_{THB} = R3 \parallel R4$	

Table 23-2: Calculated Circuit Values for Example Bridge Circuit with Variable Load

RL	$R_T = R_{THA} + RL + R_{THB}$	$I_{RL} = (V_{THA} - V_{THB}) / R_T$	$V_{RL} = I_{RL} \times RL$
0 kΩ			
1 kΩ			
2 kΩ			
3 kΩ			

RL	$R_T = R_{THA} + RL + R_{THB}$	$I_{RL} = (V_{THA} - V_{THB}) / R_T$	$V_{RL} = I_{RL} \times RL$
4 kΩ			
5 kΩ			
6 kΩ			
7 kΩ			
8 kΩ			
9 kΩ			
10 kΩ			

23.2.2 Finding I_{RL} for a Bridge Leg Variable Resistance

Refer to the circuit shown in Figure 23-7, which shows a bridge circuit with a fixed load resistor **RL** connected between nodes A and B and a variable resistance for **R4**.

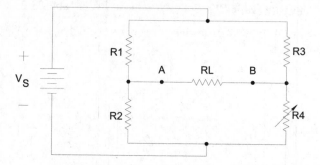

Figure 23-7: Bridge Circuit with Variable Leg Resistance and Fixed Load

You could use branch current, loop current, or node voltage analysis to calculate the current through **RL**. This example will use the loop current analysis method as shown in Figure 23-8.

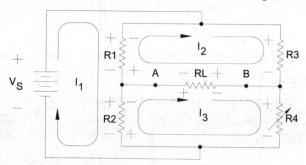

Figure 23-8: Loop Current Analysis Method for Bridge Circuit

The loop equations for this circuit are

$$-V_S + (+R1 \times I_1) + (+R2 \times I_1) + (-R1 \times I_2) + (-R2 \times I_3) = 0$$

$$(+R1 \times I_2) + (+R3 \times I_2) + (+RL \times I_2) + (-R1 \times I_1) + (-RL \times I_3) = 0$$

$$(+R2 \times I_3) + (+RL \times I_3) + (+R4 \times I_3) + (-R2 \times I_1) + (-RL \times R_2) = 0$$

Simplify and rearrange these equations to give

$$[(R1 + R2) \times I_1] + (-R1 \times I_2) + (-R2 \times I_3) = V_S$$

$$(-R1 \times I_1) + [(R1 + R3 + RL) \times I_2] + (-RL \times I_3) = 0$$

$$(-R2 \times I_1) + (-RL \times R_2) + [(R2 + RL + R4) \times I_3] = 0$$

The matrix expression for this is

$$\begin{vmatrix} R1 + R2 & -R1 & -R2 \\ -R1 & R1 + R3 + RL & -RL \\ -R2 & -RL & R2 + RL + R4 \end{vmatrix} \cdot \begin{vmatrix} I1 \\ I2 \\ I3 \end{vmatrix} = \begin{vmatrix} V_S \\ 0\ V \\ 0\ V \end{vmatrix}$$

From the circuit shown in Figure 23-8 the current through the fixed resistance RL is $I_{AB} = I_3 - I_2$.

Solving this system for each value of $R4$ may seem tedious at first, but note that the value of $R4$ affects only the shaded term in the matrix. Consequently you need change only this matrix value to solve the system when the value of $R4$ changes.

For the circuit of Figure 23-9, calculate the loop current values and value of I_{AB} for $R4 = 5$ kΩ. Record your answers in Table 23-3.

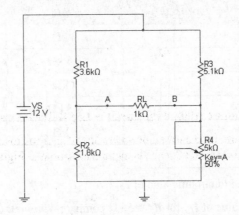

Figure 23-9: Example Bridge Circuit with Variable Leg Resistance

Table 23-3: Calculated Current Values for Figure 23-9

Circuit Value	Calculated	Circuit Value	Calculated
I_1		I_3	
I_2		$I_{AB} = I_3 - I_2$	

23.3 Design Verification

23.3.1 Finding V_{OUT} for a Variable Load Resistor

Open the circuit file Ex23-01 and simulate the circuit. Measure and record V_{RL} for each of the values of RL in Table 23-4.

Table 23-4: Measured Output Voltages for Bridge Circuit with Variable Load

RL	Measured V_{RL}	RL	Measured V_{RL}
0 kΩ		6 kΩ	
1 kΩ		7 kΩ	
2 kΩ		8 kΩ	
3 kΩ		9 kΩ	
4 kΩ		10 kΩ	
5 kΩ			

Do your measured results in Table 23-4 agree with your calculated values for V_{RL} in Table 23-2?

23.3.2 Finding I_{RL} for a Bridge Leg Variable Resistance

Open the circuit file Ex23-02. Measure and record the value of I_{RL} for each of the values of *R4* shown in Table 23-5.

Table 23-5: Measured Load Current for Bridge Circuit with Variable Leg Resistance

R4	Measured I_{RL}	R4	Measured I_{RL}
0 kΩ		3 kΩ	
500 Ω		3.5 kΩ	
1 kΩ		4 kΩ	
1.5 kΩ		4.5 kΩ	
2 kΩ		5 kΩ	
2.5 kΩ			

Which value of *R4* gives the minimum value of I_{RL}?

How does your measured value of I_{RL} for *R4* = 5 kΩ compare with your calculated value in Table 23-3?

23.4 Application Exercise

Your measured values in Table 23-5 should have shown that some value of *R4* produced a minimum value for current I_{RL}. The Wheatstone bridge operates on this principle to find the value of an unknown resistance *Rx* from fixed resistors *R1* and *R2* and a variable resistance *Rv*. Refer to Figure 23-10.

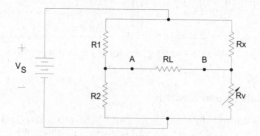

Figure 23-10: Wheatstone Bridge

The voltage across and current through RL will be 0 (that is, the bridge is balanced) when $V_A = V_B$. From Figure 23-5 this will be true when $V_S \times [R2/(R1 + R2)] = V_S \times [Rv/(Rx + Rv)]$ so

$$V_S \times [R2 / (R1 + R2)] = V_S \times [Rv / (Rx + Rv)]$$

$$R2 / (R1 + R2) = Rv / (Rx + Rv)$$

$$R2 \times (Rx + Rv) = Rv \times (R1 + R2)$$

$$(R2 \times Rx) + (R2 \times Rv) = (R1 \times Rv) + (R2 \times Rv)$$

$$R2 \times Rx = R1 \times Rv$$

$$Rx = Rv \times (R1 / R2)$$

The resistors $R1$ and $R2$ set the sensitivity of the bridge. If $R1 = R2$ then the bridge will balance when $Rv = Rx$. This means that if you can change the value of Rv in 1 Ω increments then you can measure Rx to within 1 Ω. If, however, $R2 = 100 \times R1$ then the bridge will balance when $Rv = 100 \times Rx$. This means that if you can change the value of Rv in 1 Ω increments then you can measure Rx to within 0.01 Ω.

Open the circuit Pr23-01, for which $R2 = R1$. Simulate the circuit and change the value of $Rv1$ to each of the values in Table 23-6. Record your calculated value for $Rx1$ and the measured value of $VOUT1$ for each value of $Rv1$.

Open the circuit Pr23-02, for which $R2 = 10 \times R1$. Simulate the circuit and change the value of $Rv2$ to each of the values in Table 23-6. Record your calculated value for $Rx2$ and measured value of $VOUT2$ for each value for $Rv2$.

Table 23-6: Measured Wheatstone Bridge Values

$Rv1$	$Rx1 = Rv1$	$VOUT1$	$Rv2$	$Rx2 = Rv2/10$	$VOUT2$
490 Ω			4940 Ω		
491 Ω			4941 Ω		
492 Ω			4942 Ω		
493 Ω			4943 Ω		
494 Ω			4944 Ω		
495 Ω			4945 Ω		
496 Ω			4946 Ω		
497 Ω			4947 Ω		
498 Ω			4948 Ω		
499 Ω			4949 Ω		
500 Ω			4950 Ω		

Which bridge gives the most accurate measurement for Rx and lowest value of $VOUT$?

What is the apparent tradeoff?

23.5 Section Summary

This section discussed resistive bridge circuits and presented methods for analyzing the bridge currents and voltages for varying load and leg resistances. This section also covered basic operation of the Wheatstone bridge for determining the value of an unknown resistance and demonstrated the scaling effects of $R1$ and $R2$ on the variable resistance Rv.

24. Delta and Wye Circuits

24.1 Introduction

Delta and wye circuits are three-terminal circuits, so-named because the delta circuit resembles the Greek letter delta (Δ) and the wye circuit resembles the Latin letter wye (Y). Other names for these circuits are pi and tee circuits, because they can also resemble the Greek letter pi (Π) and Latin letter tee (T). Refer to Figure 24-1.

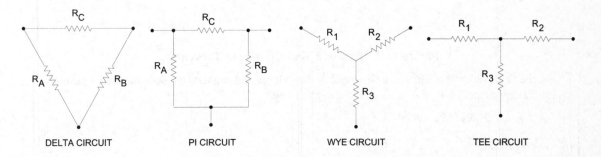

Figure 24-1: Delta (Pi) and Wye (Tee) Circuits

The primary applications for delta and wye circuits are three-phase ac circuits, such as power distribution circuits and electric motors, but the delta and wye configurations can appear in other electrical circuits. For example, *R1*, *R3*, and *RL* (or *RL*, *R2*, and *R4*) in Figure 24-2 form a delta configuration.

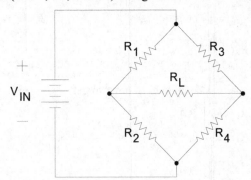

Figure 24-2: Bridge Circuit with Load Resistor

One important feature of delta and wye circuits is that every delta circuit has a wye circuit equivalent and vice versa. You will find it convenient at times to convert from one form to the other to simplify circuit analysis.

In this section you will:

- Derive the equations for converting a delta circuit to its wye circuit equivalent and vice versa.

- Determine the circuit equivalents for delta and wye circuits.

- Use the Multisim software to verify the validity of the delta and wye circuit equivalents.

24.2 Pre-Lab

To derive the circuit conversion, begin with the Thevenin equivalent for both circuit configurations as shown in Figure 24-3.

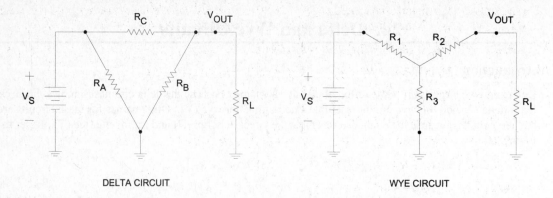

DELTA CIRCUIT   WYE CIRCUIT

Figure 24-3: Delta and Wye Circuits to Thevenize

For the delta and wye circuits to be equivalent, the voltage and resistance values for their Thevenin equivalents must also be the same. For the delta circuit

$$V_{TH\Delta} = V_S R_B / (R_B + R_C)]$$

$$R_{TH\Delta} = R_B \parallel R_C$$

$$= 1 / (1/R_B + 1/R_C)$$

$$= (R_B R_C) / (R_B + R_C)$$

For the wye circuit

$$V_{THY} = V_S R_3 / (R_1 + R_3)]$$

$$R_{THY} = R_2 + R_1 \parallel R_3 = R_2 + 1 / (1/R_1 + 1/R_3)$$

$$= R_2 + (R_1 R_3) / (R_1 + R_3)$$

$$= (R_1 R_2 + R_1 R_3 + R_2 R_3) / (R_1 + R_3)$$

Set the Thevenin voltage equations equal to each other to get

$$V_S R_B / (R_B + R_C) = V_S R_3 / (R_1 + R_3)$$

$$R_B / (R_B + R_C) = R_3 / (R_1 + R_3)$$

$$(R_B + R_C) / R_B = (R_1 + R_3) / R_3$$

$$1 + (R_C / R_B) = 1 + (R_1 / R_3)$$

$$R_C / R_B = R_1 / R_3$$

Set the Thevenin resistance equations equal to each other to get

$$R_B R_C / (R_B + R_C) = (R_1 R_2 + R_1 R_3 + R_2 R_3) / (R_1 + R_3)$$

Because of the symmetry of the delta and wye circuits, rotating the delta and wye configurations 120° clockwise yields

$$R_A / R_C = R_3 / R_2 \text{ for the voltages.}$$

$$R_C R_A / (R_C + R_A) = (R_1 R_2 + R_1 R_3 + R_2 R_3) / (R_3 + R_2) \text{ for the resistances.}$$

Rotating the delta and wye configurations 120° further clockwise yields

$$R_B / R_A = R_2 / R_1 \text{ for the voltages.}$$

$$R_C R_A / (R_C + R_A) = (R_1 R_2 + R_1 R_3 + R_2 R_3) / (R_2 + R_1) \text{ for the resistances.}$$

For the wye-to-delta circuit conversion, begin with the original resistance equation and express the values of R_B in terms of R_C on the left-hand side. From the voltage equations $R_B = R_C R_3 / R_1$ so

$$R_B R_C / (R_B + R_C) = (R_1 R_2 + R_1 R_3 + R_2 R_3) / (R_1 + R_3)$$

$$[(R_C R_3 / R_1) R_C] / [(R_C R_3 / R_1) + R_C] = (R_1 R_2 + R_1 R_3 + R_2 R_3) / (R_1 + R_3)$$

$$[(R_C R_3 / R_1) R_C] / [R_C(1 + R_3 / R_1)] = (R_1 R_2 + R_1 R_3 + R_2 R_3) / (R_1 + R_3)$$

$$(R_C R_3 / R_1) / (1 + R_3 / R_1) = (R_1 R_2 + R_1 R_3 + R_2 R_3) / (R_1 + R_3)$$

$$(R_C R_3 / R_1) / [(R_1 + R_3) / R_1] = (R_1 R_2 + R_1 R_3 + R_2 R_3) / (R_1 + R_3)$$

$$(R_C R_3 / R_1) = [(R_1 + R_3) / R_1][(R_1 R_2 + R_1 R_3 + R_2 R_3) / (R_1 + R_3)]$$

$$(R_C R_3 / R_1) = (R_1 R_2 + R_1 R_3 + R_2 R_3) / R_1$$

$$R_C = (R_1 / R_3)[(R_1 R_2 + R_1 R_3 + R_2 R_3) / R_1]$$

$$R_C = (R_1 R_2 + R_1 R_3 + R_2 R_3) / R_3$$

You can use the same method for the other resistance and voltage equations to solve for the values of R_A and R_B, but it again readily follows from the symmetry of the delta and wye circuits that $R_A = (R_1 R_2 + R_1 R_3 + R_2 R_3) / R_2$ and $R_B = (R_1 R_2 + R_1 R_3 + R_2 R_3) / R_1$.

For the delta-to-wye conversion, start with the original resistance equation and express the values of R_2 and R_3 in terms of R_1 in the right-hand expression. From the Thevenin voltage equations $R_2 = R_1 R_B / R_A$ and $R_3 = R_1 R_B / R_C$ so

$$R_B R_C / (R_B + R_C) = (R_1 R_2 + R_1 R_3 + R_2 R_3) / (R_1 + R_3)$$

$$R_B R_C / (R_B + R_C) = [(R_1^2 R_B / R_A) + (R_1^2 R_B / R_C) + (R_1^2 R_B^2 / R_A R_C)] / (R_1 + R_1 R_B / R_C)$$

$$R_B R_C / (R_B + R_C) = [R_1^2 R_B(1/R_A + 1/R_C + R_B / R_A R_C)] / [R_1(1 + R_B) / R_C)]$$

$$R_B R_C / (R_B + R_C) = [R_1 R_B(1/R_A + 1/R_C + R_B / R_A R_C)] / [(R_B + R_C) / R_C]$$

$$[R_B R_C / (R_B + R_C)][(R_B + R_C) / R_C] = R_1 R_B(1/R_A + 1/R_C + R_B / R_A R_C)$$

$$R_B = R_1 R_B(R_C / R_A R_C + R_A / R_A R_C + R_B / R_A R_C)$$

$$R_B = [R_1 R_B(R_A + R_B + R_C)] / (R_A R_C)$$

$$R_1 = R_A R_B R_C / [R_B(R_A + R_B + R_C)]$$

$$R_1 = R_A R_C / (R_A + R_B + R_C)$$

Although you can use the same method for the other resistance and voltage equations to solve for R_2 and R_3, it readily follows from the symmetry of the delta and wye circuits that $R_2 = R_B R_C / (R_A + R_B + R_C)$ and $R_3 = R_A R_B / (R_A + R_B + R_C)$.

Use the delta-wye conversion equations to complete the values for Table 24-1.

Table 24-1: Calculated Delta and Wye Circuit Conversion Values

Delta Circuit	Wye Circuit	Wye Circuit	Delta Circuit
$R_A = 100\ \Omega$	$R_1 =$	$R_1 = 100\ \Omega$	$R_A =$
$R_B = 200\ \Omega$	$R_2 =$	$R_2 = 200\ \Omega$	$R_B =$
$R_C = 300\ \Omega$	$R_3 =$	$R_3 = 300\ \Omega$	$R_C =$

24.3 Design Verification

Open the circuit Ex24-01 and change the values of the wye circuit resistors to your calculated values in Table 24-1. Record the resistance values indicated in Table 24-2.

Table 24-2: Measured Values for Calculated Wye Circuit Values

Resistance	Delta Circuit	Wye Circuit
Node X to Node Y		
Node Y to Node Z		
Node Z to Node X		

Do the corresponding delta and wye resistance values match?

Open the circuit Ex24-02 and change the values of the delta circuit resistors to your calculated values in Table 24-3. Record the resistance values indicated in Table 24-3.

Table 24-3: Measured Values for Calculated Delta Circuit Values

Resistance	Wye Circuit	Delta Circuit
Node X to Node Y		
Node Y to Node Z		
Node Z to Node X		

Do the delta and wye resistance values match?

24.4 Application Exercise

The delta-to-wye conversion equation can be useful for analyzing bridge circuits. Refer to the bridge circuit of Figure 24-4.

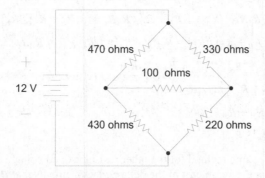

Figure 24-4: Bridge Circuit

Calculate the values required to replace the inverted delta configuration of the top half of the bridge with an inverted wye configuration and record your values in Table 24-4.

Table 24-4: Calculated Inverted Wye Resistances

Delta Resistance	Value	Wye Resistance	Value
R_A	330 Ω	R_1	
R_B	470 Ω	R_2	
R_C	100 Ω	R_3	

Open the circuit Pr24-01. Change the resistance values of the converted bridge circuit to your calculated values and simulate the circuit. Measure and record the values indicated in Table 24-5.

Table 24-5: Measured Bridge Circuit Node Values

Circuit Value	Original Bridge	Converted Bridge
I_T		
V_{AB}		
V_{AC}		
V_{BC}		
V_{BD}		
V_{CD}		

Do the node voltages for both configurations match?

Calculate the values required to replace the delta configuration of the bottom half of the bridge with a wye configuration and record your values in Table 24-6.

Table 24-6: Calculated Wye Circuit Resistances

Delta Resistance	Value	Wye Resistance	Value
R_A	430 Ω	R_1	
R_B	220 Ω	R_2	
R_C	100 Ω	R_3	

Open the circuit Pr24-02. Change the resistance values of the converted bridge circuit to your calculated values and simulate the circuit. Measure and record the voltages indicated in Table 24-7.

Table 24-7: Measured Bridge Circuit Node Values

Circuit Voltage	Original Bridge	Converted Bridge
I_T		
V_{AB}		
V_{AC}		
V_{BC}		
V_{BD}		
V_{CD}		

Do the node voltages for both configurations match?

24.5 Section Summary

This section derived the equations for converting delta circuits to wye circuit equivalents and vice versa. These equations allow you to replace portions of circuits with a configuration that may be simpler to analyze than the original form. For example, you cannot directly calculate the total resistance of the bridge circuit in Figure 24-4 using the formulas for total series and parallel resistance, but you can calculate the total resistance of either of the converted bridge circuits in the Pr24-01 and Pr24-02 application exercise files with little difficulty.

25. The Oscilloscope

25.1 Introduction

Up to now you have used the multimeter for your circuit measurements. The multimeter is ideal for making many types of precise circuit measurements, but its intent is to measure values that are static (i.e., values that do not significantly change over time). The instrument for measuring dynamic, or time-varying, values is the oscilloscope. An oscilloscope is generally less accurate than a digital multimeter and limited to voltage measurements but can display rapidly changing signals. Most oscilloscopes have multiple channels so that you can compare or combine signals, measure delays between events, and use an event on one channel to initiate the capture of information on another. Other common features of modern oscilloscopes include measurement cursors, digital time and amplitude displays, automatic detection of maximum and minimum voltages, provisions to upload waveforms to removable media or computers, and on-line help screens.

The Multisim software offers several varieties of oscilloscopes. The generic two- and four-channel oscilloscopes provide basic features that are common to most oscilloscopes. The other oscilloscopes emulate the form, features, and functions of actual oscilloscope models. Although the latter are excellent tools for learning to use the actual oscilloscopes, this section will use the two-channel generic oscilloscope to present the basic concepts and features that are common to all oscilloscopes.

In this section you will

- Learn the typical features of an oscilloscope.

- Learn how to read an oscilloscope display.

- Learn how to use the amplitude and timebase controls.

- Use the oscilloscope to make some basic circuit measurements.

25.2 Pre-Lab

25.2.1 Accessing the Oscilloscope

To access the two-channel oscilloscope, click the "Oscilloscope" tool (refer to Figure 25-1) in the Instruments toolbar.

Figure 25-1: Oscilloscope Tool

Figure 25-2 shows the minimized view of the two-channel oscilloscope for the Multisim 10 software, although the Multisim 9 oscilloscope is similar. The oscilloscope has three circuit connections. The two connections on the bottom of the oscilloscope are for the A channel (left) and B channel (right) signal and ground. The connections on the right of the oscilloscope are for the external trigger signal and ground.

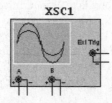

Figure 25-2: Minimized 2-Channel Oscilloscope View

Figure 25-3 shows the enlarged view of the two-channel oscilloscope with a sample display. The display is that for the Multisim 10 oscilloscope, although the Multisim 9 oscilloscope is similar.

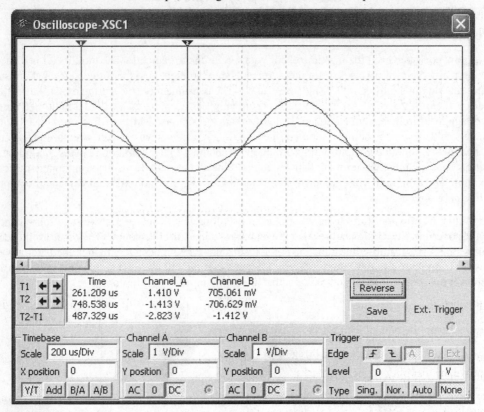

Figure 25-3: Enlarged Two-Channel Oscilloscope View

The oscilloscope controls consist of the following six sections:

1) Graphical display

2) Display controls

3) Timebase controls

4) Channel A controls

5) Channel B controls

6) Trigger controls

The following sections discuss each of the oscilloscope control sections.

25.2.2   Graphical Display

The display occupies most of the upper portion of the enlarged oscilloscope view. This is the area in which you view the channel signals and position the reference cursors. Practical oscilloscope displays are 10 divisions wide by 8 divisions high, but the Multisim generic oscilloscope displays are 10 divisions wide by 6 divisions high. The display in Figure 25-3 shows two signals, one from Channel A and one from Channel B, and reference cursors (Multisim refers to these as "crosshairs") 1 and 2. The small triangle at the top of each cursor identifies the number of the cursor.

The reference cursors provide amplitude and time information for a specific point on the channel signals. The cursors are normally located at the far right and left sides of the display. To directly position a cursor, click and drag the cursor to the desired position in the display and release the left mouse button. You can also right-click the cursor and select a specific time or amplitude value for the cursor.

Directly beneath the graphical display is a scroll bar. If the collected data extends beyond one screen, you can use the scroll bar to examine parts of the signal that are not on the screen.

25.2.3 Display Controls

The display controls are just below the oscilloscope display area. This section allows you to position the cursors, view the amplitude and time information for the channel signals at the cursor positions, and select the color of the display background. Refer to Figure 25-4.

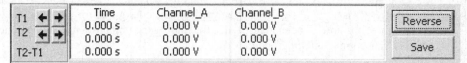

Figure 25-4: Oscilloscope Display Controls

The ← and → buttons to the right of the **T1** and **T2** labels adjust the position of reference cursors 1 and 2, respectively. The information associated with reference cursors 1 and 2 is in the window to the right of the **T1** and **T2** buttons, respectively. The amplitude and time information associated with each cursor will update as you adjust the position of the cursors.

The information in the window to the right of the **T2-T1** label is the difference in amplitude and time between reference cursor 2 and reference cursor 1. This feature allows you to easily calculate time delays, signal periods, peak-to-peak amplitudes, and other differential data.

The **Reverse** button allows you to change the background of the graphical display to black or white to improve the visibility of the signals.

The **Save** button allows you to save the graphical display as a list of time and amplitude data points in a scope display (.SCP), LabVIEW measurement (.LVM), or TDM file format so that you can import the data into other applications. The .SCP and .LVM files are in text format that you can open with a number of text editors, word processors, and spreadsheets. In addition, the Multisim program's Grapher utility (under the **View** menu) can open the .SCP file and display it as a graphic that you can view, save, and print. The .TDM file format is in a binary file format that is compatible with National Instruments' DIAdem data management and analysis software.

25.2.4 Timebase Controls

The controls in the **Timebase** section allow you to adjust the horizontal position and scale of the display and select the format of the display. Refer to Figure 25-5.

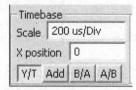

Figure 25-5: Oscilloscope Timebase Controls

The **Scale** value specifies how much time each horizontal division represents. The time settings use a 1-2-5 progression so that each setting is about twice that of the previous setting. For example, the setting before the 200 μs/Div setting is 100 μs/Div and the setting after it is 500 μs/Div. You can use the up and down scroll buttons to set this value from 1 ps (10^{-12} seconds) per divison to 1000 Ts (10^{12} seconds) per division. Unless you have nothing else to do for a while should avoid using the 1000 Ts/div setting, as each horizontal division at this setting equals approximately 31.7 million years.

The **X position** value allows you to manual shift the display in 0.1-division increments or use the scroll arrows to shift the display in 0.2-division increments to the left or right. This allows you to better align the display with a specific point on the horizontal axis.

The four buttons at the bottom of the **Timebase** section allow you to choose the format of the graphical display.

- The **Y/T** button configures the oscilloscope to display the Channel A and Channel B signals separately with the vertical axis configured for volts and the horizontal axis configured for time. This is the typical operating mode of oscilloscopes.

- The **Add** button configures the oscilloscope to add the Channel A and Channel B signals and display the result as a single signal with the vertical axis configured for volts and the horizontal axis configured for time. The mode is useful for finding the voltage across a component that has no direct connection to circuit ground.

- The **B/A** button configures the oscilloscope to plot the Channel A signal against the horizontal axis and the Channel B signal against the vertical axis to create a two-dimensional plot called a Lissajous figure. This is a convenient display mode for determining the relative amplitude, frequency, and phase of two signals. Both the vertical and horizontal axes are configured for amplitude although the oscilloscope shows no units.

- The function of the **A/B** button is similar to the **B/A** button, except that the oscilloscope plots the Channel B signal against the horizontal axis and the Channel A signal against the vertical axis.

Note that the **Scale** and **X position** controls will work only with the **Y/T** and **Add** modes.

25.2.5   Channel A Controls

The controls in the **Channel A** section allow you to adjust the vertical position and scale of the Channel A signal. Refer to Figure 25-6.

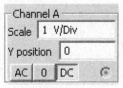

Figure 25-6: Oscilloscope Channel A Controls

The **Scale** value specifies how many volts each vertical division represents. The voltage settings use a 1-2-5 progression so that each setting is about twice the value of the previous setting. For example, the setting before the 1 V/Div setting is 500 mV/Div, and the setting after it is 2 V/Div. You can use the up and down scroll buttons to set this value from 1 pV per division to 1000 TV per division. Just for reference, 1000 TV is enough electrical potential to generate a lightning bolt 189,000 miles long. If you plan to regularly measure voltages on this order of magnitude be sure to observe adequate safety precautions.

The **Y position** value allows you to manual shift the display in 0.1-division increments or use the scroll arrows to shift the display in 0.2-division increments up or down. This allows you to separate the Channel A and B signals for better viewing or to compensate for some unwanted dc offset in the signal.

The **AC**, **0**, and **DC** buttons determine the signal coupling for the channel.

- The **AC** button removes any dc offset from the signal, so that Channel A couples (allows in) only the ac portion of the signal into the oscilloscope.

- The **0** button connects Channel A directly to ground. This is useful if you want to determine a 0 V reference for a signal on Channel A or if you wish to view only the B channel when the oscilloscope is in the **Add** mode.

- The **DC** button couples both the ac and dc components of the signal on Channel A into the oscilloscope. You will often require this coupling when you are viewing low-frequency signals so that the oscilloscope does not attenuate the signal.

25.2.6   Channel B Controls

The controls in the **Channel B** section are identical to those in the Channel A section, with the addition of one more button. Refer to Figure 25-7.

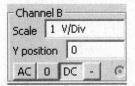

Figure 25-7: Oscilloscope Channel B Controls

The extra button, marked with a "–", inverts the signal on Channel B. You typically use this button when you wish to find the difference between the Channel A and B signals. To do this, select the **Add** mode and activate the Channel B "–" button. Because this will invert the Channel B signal, the oscilloscope will display Channel A – Channel B, rather than Channel A + Channel B.

25.2.7   Trigger Controls

The controls in the **Trigger** section determine the conditions that will trigger the oscilloscope (that is, cause the oscilloscope to display waveforms). Refer to Figure 25-8.

Figure 25-8: Oscilloscope Trigger Control

The **Edge** controls specify whether the trigger voltage must be increasing (rising edge) or decreasing (falling edge) for the oscilloscope to display the Channel A and Channel B signals. Refer to Figure 25-9, which specifies a rising edge trigger, also called a leading edge or positive edge trigger. This means that the trigger voltage must exceed the **Level** value to trigger the oscilloscope. A falling edge trigger, also called a trailing edge or negative edge trigger, means that the trigger voltage must fall below the **Level** value to trigger the oscilloscope.

Figure 25-9: Rising Edge Trigger

The **A**, **B**, and **Ext** buttons specify whether oscilloscope uses the signal on Channel A, Channel B, or External Trigger for the trigger voltage.

The **Level** value sets the voltage level for the trigger signal. You can use the scroll buttons to specify the value of the trigger signal or manually enter the value in the text box. Click in the units box and select the unit you wish to use for the trigger level from the list.

The **Type** buttons determine the type of triggering.

- The **Sing.** (single-sweep) button configures the oscilloscope to make a single sweep when the oscilloscope receives a valid trigger. After the oscilloscope completes a sweep across the screen, it should halt until you use one of the **Type** buttons to initiate a new sweep. In actuality the scope will continue to capture waveforms, although the display will continue to show the date from the first screen.

- The **Nor.** (normal) button is similar to the single-sweep button, but after the oscilloscope completes a sweep across the screen it will clear the screen and initiate a new sweep if it receives a valid trigger.

- The **Auto** (auto-trigger) button initiates a sweep whenever either of the following events occurs:

   - The oscilloscope receives a valid trigger.

   - A pre-defined amount of time has passed and the oscilloscope has not received a valid trigger.

- The **None** button specifies that there are no specific trigger conditions.

Most applications will use the auto-trigger mode, although nonperiodic signals or special conditions can require other trigger modes for best results.

25.2.8 Oscilloscope Measurement Terminology

A static value possesses only a single characteristic, namely magnitude or amplitude, that describes it. Time-varying signals have both time and amplitude characteristics to describe them. When you use an oscilloscope to observe a time-varying signal you will measure specific amplitude and time characteristics for the signal.

The most common signal you will observe in ac electronics is the sine wave. Refer to Figure 25-10.

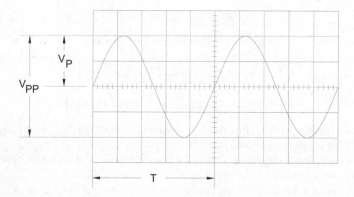

Figure 25-10: Example of Sine Wave Display

V_{PP} is the peak-to-peak voltage. The peak-to-peak voltage for a sine wave is the difference between the minimum and maximum amplitudes. V_{PP} for the sine wave in Figure 25-10 is four divisions.

V_P is the peak voltage, which is half the peak-to-peak value for a sine wave. V_P for the sine wave in Figure 25-10 is two divisions.

T is the period of the sine wave, which is the time required for one cycle of the sine wave to repeat. You will usually measure the period between consecutive positive zero-crossing points for the sine wave as shown in Figure 25-10, but you can measure the period between any two corresponding points on consecutive cycles. T for the sine wave in Figure 25-10 is five divisions.

Another characteristic of sine waves is f, the frequency. The frequency is the number of times per second that a sine wave repeats and is equal to the reciprocal of the period, $1/T$. The unit of frequency is Hertz (Hz).

25.2.9 Reading Oscilloscope Displays

To understand oscilloscope measurements you must learn to read the display. Refer to Figure 25-11.

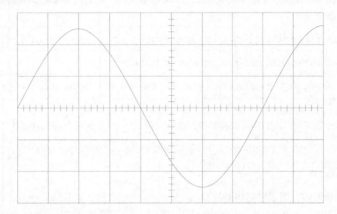

Figure 25-11: Example Oscilloscope Display

The oscilloscope display does not show numerical markings for the divisions. To determine the amplitude and period of the signal you must either

1) use the timebase and channel scale settings to convert the number of vertical and horizontal divisions into volts and seconds, or

2) use the cursors so that the Multisim program displays the measurement of interest in the display controls window.

For the waveform of Figure 25-11, determine and record the values of V_{PP}, V_P, T, and f for each of the timebase and channel settings in Table 25-1.

Table 25-1: Calculated Waveform Values for Figure 25-11

Waveform Value	Timebase = 5 µs/Div Channel = 200 µV/Div	Timebase = 200 µs/Div Channel = 50 mV/Div	Timebase = 1 ms/Div Channel = 10 V/Div
V_{PP}			
$V_P = V_{PP} / 2$			
T			
$f = 1/T$			

25.2.10 Determining the Oscilloscope Timebase and Voltage Scale Settings

If you know the approximate frequency and amplitude of a signal that you wish to measure, it is a good idea for you to know the oscilloscope timebase and channel scale settings that you will use to measure the signal. This provides a check on the signal values that you expect to measure. Ideally you would like one cycle to occupy the entire oscilloscope display so that the peak-to-peak signal amplitude is six divisions high, and the period is ten divisions wide. Practically, however, you must round up to the nearest 1-2-5 scale settings that will show the full signal amplitude and period.

For each of the signal amplitudes and frequencies listed in Table 25-2, calculate and record the ideal and practical (1-2-5 progression) oscilloscope timebase and channel scale settings.

Table 25-2: Calculated Oscilloscope Scale Settings

Waveform Value	$Vp = 200$ mV $f = 250$ kHz	$Vp = 10$ V $f = 400$ Hz	$Vp = 189$ V $f = 60$ kHz
$V_{PP} = 2 \times V_P$			
Ideal Channel Scale $= V_{PP} / 6$			
Practical Channel Scale			
$T = 1/f$			
Ideal Timebase Scale $= T / 10$			
Practical Channel Scale			

25.3 Design Verification

Open the circuit file Ex25-01. Use the right-click menu for the ac voltage source to set the **Voltage (Pk)** and **Frequency (F)** values to your calculated values of V_P and f in Table 25-1. Simulate the circuit and verify that the oscilloscope display matches that of Figure 25-11 for the corresponding timebase and channel scale settings of Table 25-2.

Open the circuit file Ex25-02. Use the right-click menu for the ac voltage source to set the **Voltage (Pk)** and **Frequency (F)** values to the values of V_P and f in Table 25-2. Set the timebase and channel scale settings to the corresponding practical scale settings you calculated for Table 25-2 and simulate the circuit.

Do your calculated settings allow you to view one complete cycle of the waveform?

Can you decrease either the timebase or channel scale setting and still view the complete waveform?

25.4 Section Summary

This section introduced the oscilloscope and discussed the various features of the Multisim generic two-channel oscilloscope that are available in some form on practical oscilloscopes. This section also covered the practical aspects of reading the oscilloscope display to measure amplitude and frequency and determining the timebase and channel scale settings for making oscilloscope measurements. You will be using the oscilloscope extensively to work with reactive and other ac circuits.

26. DC Response of RC Circuits

26.1 Introduction

Physically a capacitor consists of two conductors, called plates, separated by an insulator, or dielectric. Functionally a capacitor stores charge on the plates and energy in an electric field between the plates.

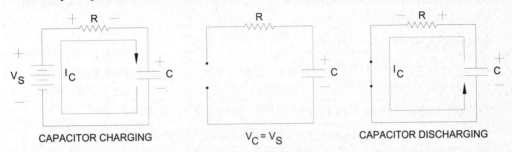

Figure 26-1: Series RC Circuit

Figure 26-1 shows a series RC circuit, so called because it consists of a resistor (R) and capacitor (C) in series. When you apply a voltage source to the circuit the source moves charge off one capacitor plate and deposits the charge on the other plate. This movement of charge is capacitor current I_C, and the separation of charge creates a potential difference, or voltage, across the capacitor. When $V_C = V_S$ no further current will flow, and the capacitor will hold its charge even if you remove the voltage source from the circuit. If a path exists between the charged capacitor plates, a discharge current will move charge from one plate back to the other plate. This current will continue until $V_C = 0$.

When a capacitor is charging or discharging, the voltage across it at time t is

$$V_C(t) = V_F + (V_0 + V_F)e^{-(t/\tau)}$$

where

$V_C(t)$ is the capacitor voltage at time t,

V_0 is the initial capacitor voltage,

V_F is the final voltage to which the capacitor is charging, and

$\tau = RC$, the RC time constant.

Two special cases are when $V_0 = 0$ V (the capacitor is initially discharged) and when $V_F = 0$ V (the capacitor is completely discharging). When $V_0 = 0$ V the equation simplifies to

$$V_C(t) = V_F - V_F e^{-(t/\tau)} = V_F(1 - e^{-(t/\tau)})$$

When $V_F = 0$ V the equation reduces to

$$V_C(t) = V_0 e^{-(t/\tau)}$$

In this section you will:

- Determine the significance of the RC time constant.
- Study the charging and discharging characteristics of RC circuits.
- Learn about the elapsed simulation time counter.
- Learn to work with the Multisim oscilloscope display cursors.
- Use the Multisim software to verify the step and pulse responses of RC circuits.

26.2 Pre-Lab

26.2.1 The RC Time Constant

The RC time constant is a special characteristic of an RC circuit. One time constant is the time it takes for the capacitor to charge (or discharge) 63.2% of the way to the final voltage. When the elapsed time t is an integral multiple of RC, the exponent indicates how many time constants N have elapsed. The RC charging and discharging equations simplify to

Charging equation: $\qquad V_C(t) = V_F(1 - e^{-N}) = V_F(1 - k) = (1 - k)V_F$

Discharging equation: $\qquad V_C(t) = V_0(e^{-N}) = V_0(k) = kV_0$

For the charging equation $(1 - k)$ is the percentage of the final voltage to which the capacitor is charging. For the discharging equation k is the percentage of initial voltage remaining on the capacitor.

Calculate the values of k and $1 - k$ for each of the values of $N = t / \tau$ in Table 26-1.

Table 26-1: RC Time Constant Calculations

$N = t / \tau$	$1 - e^{-N}$	e^{-N}	$N = t / \tau$	$1 - e^{-N}$	e^{-N}
0			5		
1			6		
2			7		
3			8		
4			9		

How many time constants N must pass before the voltage on a capacitor charges to more than 99% of its final value?

How many time constants N must pass before the voltage on a capacitor discharges to less than 1% of its initial value?

26.2.2 Series RC Charging Example

Refer to the circuit of Figure 26-2. What is the RC time constant (RC) of the circuit?

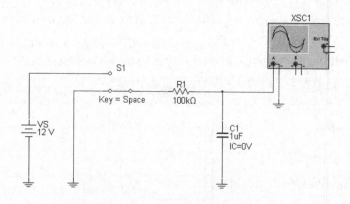

Figure 26-2: Series RC Charging Circuit

Calculate the capacitor voltage for each value of t after the switch connects to the 12V source and record your values in the "Calculated $V_C(t)$" column of Table 26-2. Assume that the initial capacitor voltage is 0 V and that the switch connects to the 12V source at t = 0.

Table 26-2: Calculated Capacitor Voltages for Figure 26-2

$N = t / \tau$	$1 - e^{-N}$	Calculated $V_C(t)$	Measured $V_C(t)$
0			
1			
2			
3			
4			
5			

26.2.3   Series RC Discharging Example

Refer to the circuit of Figure 26-3. What is the RC time constant of the circuit?

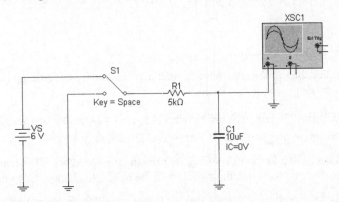

Figure 26-3: Series RC Discharging Circuit

Calculate the capacitor voltage for each value of t after the switch connects to the ground and record your values in the "Calculated $V_C(t)$" column of Table 26-3. Assume that the initial capacitor voltage is 6 V and that the switch connects to ground at t = 0.

Table 26-3: Calculated Capacitor Voltages for Figure 26-3

$N = t / \tau$	e^{-N}	Calculated $V_C(t)$	Measured $V_C(t)$
0			
1			
2			
3			
4			
5			

26.3 Design Verification

26.3.1 The Simulation Status Bar

This design verification uses the simulation status bar at the bottom of the Multisim interface. Refer to Figure 26-4.

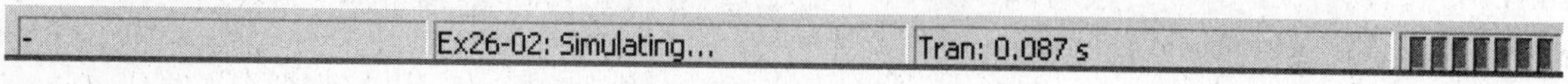

Figure 26-4: Simulation Status Bar

Although you may not have noticed it before, the Multisim program displays information in this area whenever you simulate a circuit.

The "**<Filename>: Simulating...**" message indicates that Multisim is currently simulating the indicated circuit file. When Multisim displays this message you cannot modify the circuit file.

The "**Tran: <Time>**" message indicates the elapsed circuit simulation time for the Multisim program's transient analysis. Note that this time corresponds to the elapsed *circuit simulation* time, and not the elapsed *real* time (i.e., how long the program has actually been running).

The green bars indicate that the simulation is running. This is useful for you to determine whether the elapsed time indicator is changing too slowly for you to notice or whether the program has locked up.

26.3.2 Series RC Charging Example

1. Open the circuit file Ex26-01.

2. Double-click the oscilloscope to enlarge it. Click and drag the enlarged view to another location if it obstructs your view of the circuit.

3. Click the **Run** switch to start the simulation.

4. Allow the simulation to run for 0.100 second (1 division).

5. Connect the switch to the 12V source by pressing the space bar. If you are using the Multisim 10 software, you can also click the switch with the mouse to change the switch setting.

6. Let the simulation run for at least another 0.600 second (6 divisions) after connecting the switch to the 12V source.

7. Click the **Run** switch to stop the simulation.

8. Use the scroll bar to move the oscilloscope display so that the start of the charging curve is as far to the left as possible, but still visible.

9. Right-click Cursor 1 to open the cursor right-click menu.

10. Select "Set Y_Value =>" from the menu. Purely numeric values will specify seconds for X values and volts for Y values. "T", "G", "M", "k", "m", "u", "n", and "p" after numbers indicate tera-, giga-, mega-, kilo-, milli-, micro-, nano-, and pico- suffixes. For example, an X-value of "1.500m" indicates 1.5 milliseconds and a Y-value of "2.5k" indicates 2.5 kilovolts.

11. Enter the value "0.001" or "1m" and click the **OK** button. The Multisim program will position the cursor at the first data measurement to the right of the cursor that is equal to 1 mV. This will be very near the start of the charging curve, corresponding to $t / \tau = 0$.

12. Record the "Channel A" value for T1 in the "Measured $V_C(t)$" column of Table 26-2 for the "$N = t / \tau$" value of 0.

13. Right-click Cursor 2 to open the cursor right-click menu.

14. Select "Set X_Value" from the menu.

15. Enter a value equal to T1 + τ so that the "Time" value for T2–T1 equals one time constant for the circuit. This will place Cursor 2 one time constant to the right of Cursor 1.

16. Record the Channel A value for T2 in the "Measured $V_C(t)$" column of Table 26-2 for the "$N = t / \tau$" value of 1.

17. Repeat Steps 13 through 16 using appropriate X values for the remaining values of "$N = t / \tau$" in Table 26-2.

26.3.3 Series RC Discharging Example

1. Open the circuit file Ex26-02.

2. Double-click the oscilloscope to enlarge it. Click and drag the enlarged view to another location if it obstructs your view of the circuit.

3. Click the **Run** switch to start the simulation.

4. Allow the simulation to run for 0.100 second (2 divisions).

5. Connect the switch to the ground by pressing the space bar. If you are using the Multisim 10 software, you can also click the switch with the mouse to change the switch setting.

6. Allow the simulation to run for at least 0.300 second (6 divisions) after you connect the switch to ground.

7. Click the **Run** switch to stop the simulation.

8. Use the scroll bar to move the oscilloscope display so that the start of the discharging curve is as far to the left as possible, but still visible.

9. Right-click Cursor 1 to open the cursor right-click menu.

10. Select "Set Y_Value =>" from the menu.

11. Enter the value "5.999" and click OK. The Multisim program will position the cursor at the first data measurement to the right of the cursor that is equal to 5.999 V. This will be very near the start of the discharging curve, corresponding to t / τ = 0.

12. Record the "Channel A" value for T1 in the "Measured $V_C(t)$" column of Table 26-3 for the "$N = t / \tau$" value of 0.

13. Right-click Cursor 2 to open the cursor right-click menu.

14. Select "Set X_Value" from the menu.

15. Enter a value equal to T1 + τ so that the "Time" value for T2–T1 equals one time constant for the circuit. This will place Cursor 2 one time constant to the right of Cursor 1.

16. Record the Channel A value for T2 in the "Measured $V_C(t)$" column of Table 26-3 for the "$N = t / \tau$" value of 1.

17. Repeat Steps 13 through 16 using appropriate X values for the remaining values of "$N = t / \tau$" in Table 26-3.

26.4 Application Exercise

Not all RC circuits are simple series RC circuits. Some circuits will have more complex resistor networks. To analyze the step RC response of these circuits, you must Thevenize the resistor network to reduce the circuit to a simple series RC circuit like those you have just investigated. Refer to the circuit in Figure 26-5.

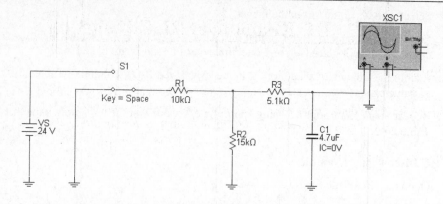

Figure 26-5: Series-Parallel RC Charging Circuit

Thevenize the circuit to the left of the capacitor and record the values for V_{TH} and R_{TH} in Table 26-4.

Table 26-4: Thevenin Values for Series-Parallel RC Circuit

V_{TH}	R_{TH}

Open the circuit file Pr26-01. Simulate a charging circuit for at least 300 milliseconds. Use the cursors to measure the capacitor voltage for each of the charging times shown in Table 26-5 and record your results in the "Original Circuit $V_C(t)$" column.

Table 26-5: Series-Parallel RC Circuit Charging Measurements

Charging Time	Original Circuit $V_C(t)$	Thevenin Circuit $V_C(t)$
0 ms		
50 ms		
100 ms		
150 ms		
200 ms		
250 ms		
300 ms		

Now, open the circuit Pr26-02. Change the **VTH** and **RTH** values for the circuit to your calculated values in Table 26-4 and simulate a charging circuit for at least 300 milliseconds. Use the cursors to obtain capacitor voltage measurements for each of the charging curve times shown in Table 26-5 and record your results in the "Thevenin Circuit $V_C(t)$" column. Do capacitor voltages agree for the original and Thevenin circuits?

26.5 Section Summary

This section explored the charging and discharging characteristics of RC circuits and discussed the use of the Multisim oscilloscope cursors for making precise measurements. Although not every RC circuit is a simple series RC circuit, you can Thevenize many complex circuits to obtain a series RC equivalent.

27. DC Response of RL Circuits

27.1 Introduction

An inductor is any device whose primary purpose is to create an electromagnetic field when current flows through it. A typical inductor physically consists of a coil of wire wound about some core material. An inductor opposes any change in the current flowing through it by developing a voltage that opposes the change. This effect on current is why inductors are referred to as "chokes" in some applications. Another term for an inductor is "coil."

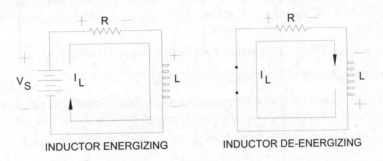

INDUCTOR ENERGIZING   INDUCTOR DE-ENERGIZING

Figure 27-1: Series RL Circuit

Figure 27-1 shows a series RL circuit, so called because it consists of a resistor (R) and inductor (L) in series. When V_S is first applied to the circuit $I_L = 0$, but as current begins to flow the inductor stores energy in an electromagnetic field that in turn develops a voltage V_L to oppose the increase in current. As I_L continues to increase its rate of change decreases so that the induced voltage V_L decreases. Eventually I_L reaches a maximum value of V_S / R and $V_L = 0$ V, at which time the inductor is fully energized and the energy in the electromagnetic field is maximum. If now V_S goes to 0 V so that the current attempts to decrease, the inductor will attempt to maintain the current value. The electromagnetic field will return energy to the circuit and develop a voltage V_L across the inductor that reinforces the flow of current. Eventually the energy in the electromagnetic field is exhausted, at which time the inductor is completely de-energized and $I_L = 0$.

When an inductor is energizing or de-energizing, the current through it at time t is

$$I_L(t) = I_F + (I_0 - I_f)\, e^{-[t/\tau]} = I_F + (I_0 - I_F)e^{-(t/\tau)}$$

where

> $I_L(t)$ is the inductor current at time t,
>
> I_0 is the initial inductor voltage,
>
> I_F is the final current through the inductor, and
>
> $\tau = L / R$, the RL time constant.

Two special cases are when $I_0 = 0$ A (the inductor is initially de-energized) and when $I_F = 0$ A (the inductor is completely de-energizing). When $I_0 = 0$ A the equation simplifies to

$$I_L(t) = I_F - I_F\, e^{-(t/\tau)} = I_F\,(1 - e^{-(t/\tau)})$$

When $I_F = 0$ V the equation reduces to

$$I_L(t) = I_0\, e^{-(t/\tau)}$$

In this section you will:

- Study the energizing and de-energizing characteristics of RL circuits.

- Learn how to use the oscilloscope to make differential voltage measurement.

- Use the Multisim software to verify the step and pulse responses of RC circuits.

- Investigate the phenomenon of inductive kickback.

27.2 Pre-Lab

27.2.1 The RL Time Constant

The RL time constant is a special characteristic of an RL circuit. One time constant is the time it takes the energizing (or de-energizing) current to reach 63.2% of the final current. When the elapsed time t is an integral multiple of L/R, the exponent indicates how many time constants N have elapsed. The RL energizing and de-energizing equations simplify to

Energizing equation: $\quad I_L(t) = I_F (1 - e^{-N}) = I_F (1 - k) = (1 - k) I_F$

De-energizing equation: $\quad I_L(t) = I_0 (e^{-N}) = I_0 (k) = k V_0$

For the energizing equation $(1 - k)$ is the percentage of the final current through the inductor. For the de-energizing equation k is the percentage of initial current provided by the inductor.

Calculate the values of k and $1 - k$ for each of the values of $N = t / \tau$ in Table 27-1.

Table 27-1: RC Time Constant Calculations

$N = t / \tau$	$1 - e^{-N}$	e^{-N}	$N = t / \tau$	$1 - e^{-N}$	e^{-N}
0			5		
1			6		
2			7		
3			8		
4			9		

How many time constants N must pass before the energizing current through the inductor is more than 99% of its final value?

How many time constants N must pass before the de-energizing current through the inductor is less than 1% of its initial value?

27.2.2 Series RL Energizing Circuit

Refer to the circuit of Figure 27-2.

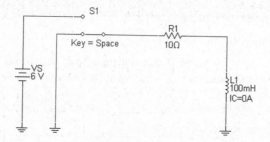

Figure 27-2: Series RL Energizing Circuit

What is the RL time constant (L/R) of the circuit?

What is I_F, the final current through the inductor?

Calculate the inductor current for each value of t after the switch connects to the 6V source and record your values in the "Calculated I_L(t)" column in Table 27-2. Assume that the initial inductor current is 0 V and that the switch connects to the 6V source at t = 0.

Table 27-2: Calculated Inductor Currents for Figure 27-2

$N = t / \tau$	$1 - e^{-N}$	Calculated I_L(t)
0		
1		
2		
3		
4		
5		

27.2.3 Series RL De-Energizing Example

Refer to the circuit of Figure 27-3. What is the RL time constant of the circuit?

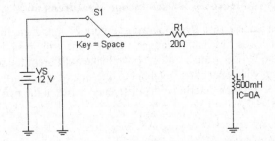

Figure 27-3: Series RL De-Energizing Circuit

Calculate the inductor current for each value of t after the switch connects to the ground and record your values in the "Calculated I_L(t)" column in Table 27-3. Assume that the initial inductor current is 600 mA and that the switch connects to ground at t = 0.

Table 27-3: Calculated Inductor Current for Figure 27-3

$N = t / RC$	e^{-N}	Calculated I_L(t)
0		
1		
2		
3		
4		
5		

27.3 Design Verification

27.3.1 Differential Oscilloscope Measurements

Using an oscilloscope to measure I_L for the circuits of Figure 27-2 and Figure 27-3 presents a problem because the oscilloscope measures voltage and not current. Because the circuits are series RL circuits, $I_L = I_R$ so you could measure V_R and calculate I_L from $I_L = I_R = V_R / R$. This, however, presents another possible problem. R is a "floating" component. In other words, neither side connects directly to ground. If the reference input for the measuring channel connects internally to the same ground as that of the circuit, the oscilloscope could short out part of the circuit. Refer to the example in Figure 27-4.

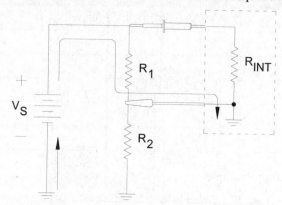

Figure 27-4: Oscilloscope Measurement

If you connect the probe leads across *R1* as shown in Figure 27-4, the oscilloscope will short out, or bypass, *R2* if the reference for the channel connects to the same ground as that of the circuit. You can avoid this sort of problem by using the oscilloscope to make a differential measurement. To make a differential measurement, you connect one channel to one side of the component of interest, the other channel to the other side, and then determine the difference between the two measurements. Refer now to the example in Figure 27-5.

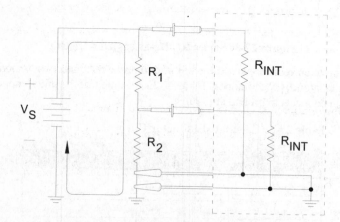

Figure 27-5: Differential Oscilloscope Measurement

As the example shows, the internal oscilloscope and circuit grounds are the same so that the differential measurement across *R1* will not short out *R2*.

Although the Multisim oscilloscope will not short out your virtual circuits, you will use the differential measurements to determine the RL energizing and de-energizing currents.

27.3.2   Series RL Energizing Example

1. Open the circuit file Ex27-01.

2. Save the file as Ex27-01_Mod.

3. Right-click the wire segment between *R1* and *L1* to open the wire segment right-click menu.

4. Select **Segment Color...** to open the **Colors** configuration window.

5. Select a different color for the wire segment. You can select the blue color shown in Figure 27-6, but any color other than red will do.

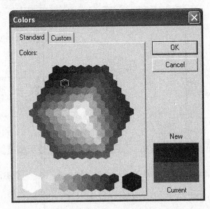

Figure 27-6: Colors Configuration Window

The reason for changing the wire segment color is that the color of the trace for a channel on the Multisim oscilloscope is the same as that for the wire segment to which the oscilloscope channel connects. Using different colors for the nodes that you will observe makes it easier to analyze the traces.

6. Connect the (+) terminal of Channel A to the wire segment between *VS* and *R1*.

7. Connect the (+) terminal of Channel B to the wire segment between *R1* and *L1*.

8. Connect the (−) terminals of Channel A and Channel B to circuit ground. If you wish, add another ground symbol to the circuit to simplify the wiring.

9. Double-click the oscilloscope to enlarge it. Move the enlarged view to another location if it obstructs your view of the circuit.

10. Change the oscilloscope channel settings to those in Table 27-4.

Table 27-4: Ex27-01 Oscilloscope Settings

Oscilloscope Settings	Channel A	Channel B
Channel Scale	2 V/Div	2 V/Div
Y position	−1	−1
Timebase scale	10 ms/Div	
X position	0	

11. Click the **Run** switch to start the simulation.

12. Allow the simulation to run for 0.020 second (2 divisions).

13. Connect the switch to the 6V source by pressing the space bar. If you are using the Multisim 10 software, you can also click the switch with the mouse to change the switch setting.

14. Let the simulation run for at least another 0.060 second (6 divisions) after connecting the switch to the 6V source.

15. Click the **Run** switch to stop the simulation. Your oscilloscope display should look like that of Figure 27-7.

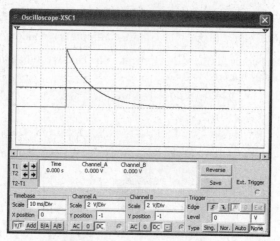

Figure 27-7: Energizing Signal Waveforms for Channels A and B

The Channel A (6V step) waveform displays the voltage at the node between *VS* and *R1* (i.e., application of the 6V source to the circuit). The Channel B waveform displays the voltage at the node between *R1* and *L1* (i.e., the voltage across *L1*). Subtracting Channel B from Channel A will give the voltage across R1. You could manually calculate this, but an easier way is to invert the signal on Channel B and add it to Channel A.

16. Change the **Timebase** display mode from "Y/T" to "Add".

17. Click the "–" button for the Channel B settings.

18. Select "Clear Instrument Data" under the Simulate menu on the menu bar.

19. Click the **Run** switch to start the simulation.

20. Allow the simulation to run for 0.020 second (2 divisions).

21. Connect the switch to the 6V source by pressing the space bar. If you are using the Multisim 10 software, you can also click the switch with the mouse to change the switch setting.

22. Let the simulation run for at least another 0.060 second (6 divisions) after connecting the switch to the 6V source.

23. Click the **Run** switch to stop the simulation. Your oscilloscope display should look like that of Figure 27-8. If it does not, repeat Steps 18 through 22.

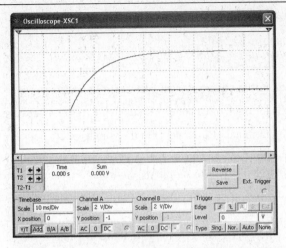

Figure 27-8: Energizing Signal Waveform for V_R

24. Right-click Cursor 1 to open the cursor right-click menu.

25. Select "Set Y_Value =>" from the menu.

26. Enter the value "0.001" or "1m" and click OK. The Multisim program will position the cursor at the first data measurement to the right of the cursor that is equal to 1 mV. This will be very near the start of the energizing curve, corresponding to $t / \tau = 0$.

27. Record the "Channel A" value for T1 in the "Measured V_R(t)" column of Table 27-5 for the "$N = t / \tau$" value of 0.

28. Right-click Cursor 2 to open the right-click menu.

29. Select "Set X_Value" from the menu.

30. Enter a value equal to T1 + τ so that the "Time" value for T2–T1 equals one time constant for the circuit. This will place Cursor 2 one time constant to the right of Cursor 1.

31. Record the Channel A value for T2 in the "Measured V_R(t)" column of Table 27-5 for the "$N = t / \tau$" value of 1.

32. Repeat Steps 28 through 31 for the remaining values of "$N = t / \tau$" in Table 27-5.

33. Calculate I_L(t) = V_R(t) / $R1$ for each of the values in Table 27-5.

Table 27-5: Measured Inductor Currents for Figure 27-2

$N = t / \tau$	Measured V_R(t)	Measured I_L(t)
0		
1		
2		
3		
4		
5		

Do the measured values in Table 27-5 match your calculated values in Table 27-2?

27.3.3   Series RL De-Energizing Example

1. Open the circuit file Ex27-02.

2. Save the file as Ex27-02_Mod.

3. Change the color of the wire segment between *R1* and *L1*.

4. Connect the (+) terminal of Channel A to the wire segment between *VS* and *R1*.

5. Connect the (+) terminal of Channel B to the wire segment between *R1* and *L1*.

6. Connect the (−) terminals of Channel A and Channel B to circuit ground. If you wish, add another ground symbol to the circuit to simplify the wiring.

7. Double click the oscilloscope to enlarge it. Move the enlarged view if it obstructs your view of the circuit.

8. Change the oscilloscope channel settings to those in Table 27-6.

Table 27-6: Ex27-02 Oscilloscope Settings

Oscilloscope Settings	Channel A	Channel B
Channel Scale	5 V/Div	5 V/Div
Y position	0	0
Timebase scale	20 ms/Div	
X position	0	

9. Click the **Run** switch to start the simulation.

10. Allow the simulation to run for 0.040 second (2 divisions).

11. Connect the switch to the 12V source by pressing the space bar. If you are using the Multisim 10 software, you can also click the switch with the mouse to change the switch setting.

12. Let the simulation run for at least another 0.120 second (6 divisions) after connecting the switch to the 6V source.

13. Click the **Run** switch to stop the simulation. Your oscilloscope display should look like that of Figure 27-9.

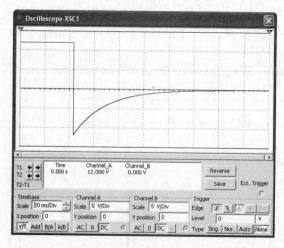

Figure 27-9: De-Energizing Signal Waveforms for Channels A and B

The Channel A (12V step) waveform displays the voltage at the node between *VS* and *R1* (i.e., application of the 12V source to the circuit). The Channel B waveform displays the voltage at the node between *R1* and *L1* (i.e., the voltage across *L1*). Subtracting Channel B from Channel A will give the voltage across *R1*.

14. Change the **Timebase** display mode from "Y/T" to "Add".

15. Click the "–" button for the Channel B settings.

16. Select "Clear Instrument Data" under the Simulate menu on the menu bar.

17. Click the **Run** switch to start the simulation.

18. Allow the simulation to run for 0.040 second (2 divisions).

19. Connect the switch to the 6V source by pressing the space bar. If you are using the Multisim 10 software, you can also click the switch with the mouse to change the switch setting.

20. Let the simulation run for at least another 0.120 second (6 divisions) after connecting the switch to the 6V source.

21. Click the **Run** switch to stop the simulation. Your oscilloscope display should look like that of Figure 27-10. If it does not, repeat Steps 16 through 20 again.

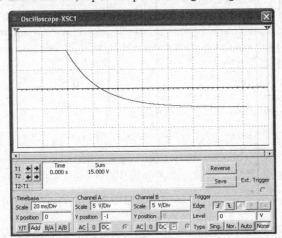

Figure 27-10: De-Energizing Signal Waveform for V_R

22. Right-click Cursor 1 to open the cursor right-click menu.

23. Select "Set Y_Value =>" from the menu.

24. Enter the value "11.999" and click OK. The Multisim program will position the cursor at the first data measurement to the right of the cursor that is equal to 11.999 V. This will be very near the start of the de-energizing curve, corresponding to t / τ = 0.

25. Record the "Channel A" value for T1 in the "Measured V_R(t)" column of Table 27-7 for the "N = t / τ" value of 0.

26. Right-click Cursor 2 to open the right-click menu.

27. Select "Set X_Value" from the menu.

28. Enter a value equal to T1 + τ so that the "Time" value for T2–T1 equals one time constant for the circuit. This will place Cursor 2 one time constant to the right of Cursor 1.

29. Record the Channel A value for T2 in the "Measured V_R(t)" column of Table 27-7 for the "N = t / τ" value of 1.

30. Repeat Steps 26 through 29 for the remaining values of "N = t / τ" in Table 27-7.

31. Calculate $I_L(t) = V_R(t) / R1$ for each of the values in Table 27-7.

Table 27-7: Measured Inductor Currents for Figure 27-10

$N = t / \tau$	Measured $V_R(t)$	Measured $I_L(t)$
0		
1		
2		
3		
4		
5		

Do your measured values in Table 27-7 match your calculated values in Table 27-3?

27.4 Application Exercise

As Figure 27-1 shows, an inductor will develop a voltage to maintain the current flowing through it if the source of the current is removed. This voltage is called "inductive kickback" and will occur whenever something interrupts current to inductive circuits, including relay coils and motor or generator windings. A potentially dangerous aspect of this phenomenon is that the induced voltage can be hundreds or thousands of volts. Refer to Figure 27-11.

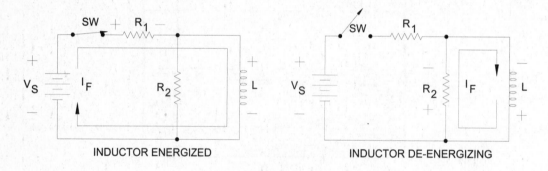

INDUCTOR ENERGIZED   INDUCTOR DE-ENERGIZING

Figure 27-11: Inductive Kickback Example

When switch *SW* closes, the current through inductor *L* increases. Since $R_L \approx 0$ the current eventually reaches its final value of $I_F = V_S / R_1$. When *SW* opens, *L* initially attempts to keep the value of the current at I_F. Because the only path for I_F is through R_2, Kirchhoff's Voltage Law and Ohm's Law require that $V_L = -I_F R_2 = -(V_S / R_1)R_2 = -V_S(R_2 / R_1)$. What is the calculated V_L for the circuit in Figure 27-12?

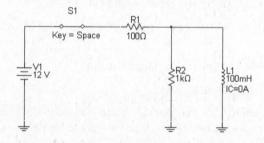

Figure 27-12: Inductive Kickback Example Circuit

1. Open the practice circuit file Pr27-01.

2. Click the **Run** switch to simulate the circuit.

3. Open the switch *S1*.

4. Click the **Run** switch to stop the simulation.

5. Use the scroll bar below the display so that the induced voltage spike is within the display area.

6. Right-click Cursor 1 and select "Go to next Y_Min =>" to find the maximum negative voltage. What is the value of the induced voltage?

7. Change the value of *R2* to 5 kΩ and repeat Steps 2 through 6. What is the value of the induced voltage?

If the resistance of the de-energizing path is much larger than the steady-state resistance (*R2*/ *R1* is very large), the magnitude of the inductive kickback voltage can be very large. Inductive kickback can damage circuitry and cause electrical arcing if the circuit design does not include proper precautions.

27.5 Section Summary

This section explored the energizing and de-energizing characteristics of RL circuits and demonstrated the use of the Multisim oscilloscope to make differential measurements. It also investigated the phenomenon of inductive kickback in inductive circuits, and showed that the induced voltage V_L could be much higher than the source voltage. While most circuit designs incorporate some means to limit this voltage, some applications make use of this induced voltage. Perhaps the most common example is the coil that fires the spark plugs in an internal combustion engine.

28. Introduction to Reactive Circuits

28.1 Introduction

Capacitors and inductors possess a frequency-dependent property called reactance. Like resistance, reactance opposes the flow of current in a circuit. Unlike resistance, reactance stores and releases energy in a circuit rather than dissipates it. In addition, reactance possesses a phase component. This means that although the frequency of the voltage across a reactive component and the current through it are the same, they are not in phase (i.e., the voltage and current will not reach their maximum, minimum, and zero-crossing values at the same time). You will learn more about the phase of reactive circuits in later sections. The symbol for reactance is X, and the unit of reactance is ohms.

Reactance can take one of two forms: capacitive and inductive. The magnitude of capacitive reactance decreases as frequency increases, whereas the magnitude of inductive reactance increases as frequency increases. The equations for calculating the magnitude of capacitive and inductive reactance in ohms at a frequency f in Hertz are

Capacitive reactance: $|X_C| = X_C = 1/[(2\pi(f)(C)]$

Inductive reactance: $|X_L| = X_L = 2\pi(f)(L)$

From the equations you can see that at dc ($f = 0$ Hz) X_C is infinite and $X_L = 0\ \Omega$. Conversely, as f becomes infinite, $X_C = 0$ and X_L is infinite. A generalization for the frequency response of capacitors and inductors is that capacitors are dc opens and ac shorts, and that inductors are dc shorts and ac opens.

In this section you will:

- Investigate the frequency characteristics of capacitors and inductors.

- Use the Multisim software to verify the equations for capacitive and inductive reactance.

- Determine the total reactance of series and parallel reactive circuits.

28.2 Pre-Lab

28.2.1 Basic Reactance

Refer to the circuit in Figure 28-1.

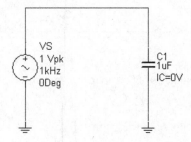

Figure 28-1: Example Capacitive Reactance Circuit

For each frequency in Table 28-1, calculate the capacitive reactance and the total current for the circuit.

Table 28-1: Calculated Values for Example Capacitive Reactance Circuit

f	$X_C = 1/(2\pi fC)$	$I_T = VS / X_C$	f	$X_C = 1/(2\pi fC)$	$I_T = VS / X_C$
10 Hz			500 Hz		
20 Hz			1 kHz		
50 Hz			2 kHz		
100 Hz			5 kHz		
200 Hz			10 kHz		

Refer to the circuit of Figure 28-2.

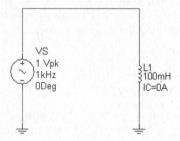

VS
1 Vpk
1kHz
0Deg

L1
100mH
IC=0A

Figure 28-2: Example Inductive Reactance Circuit

For each frequency in Table 28-2 calculate the inductive reactance and the total current for the circuit.

Table 28-2: Calculated Values for Example Inductive Reactance Circuit

f	$X_L = 2\pi fL$	$I_T = VS / X_L$	f	$X_L = 2\pi fL$	$I_T = VS / X_L$
10 Hz			500 Hz		
20 Hz			1 kHz		
50 Hz			2 kHz		
100 Hz			5 kHz		
200 Hz			10 kHz		

28.2.2 Series Reactance

The equation for the total capacitance of two capacitors in series is

$$C_T = 1 / (1/C_1 + 1/C_2) = C_1C_2/(C_1 + C_2)$$

The total reactance of two capacitors in series is therefore

$$X_T = 1/[(2\pi(f)(C_T)]$$

$$= 1/\{2\pi(f)[C_1C_2/(C_1 + C_2)]\}$$

$$= (C_1 + C_2) / [2\pi(f)(C_1C_2)]$$

$$= C_1 / [2\pi(f)(C_1C_2)] + C_2 / [2\pi(f)(C_1C_2)]$$

$$= 1 / [2\pi(f)(C_2)] + 1 / [2\pi(f)(C_1)]$$

$$= X_{C2} + X_{C1} = X_{C1} + X_{C2}$$

The equation for the total inductance of two inductors in series is

$$L_T = L_1 + L_2$$

The total reactance of two inductors in series is therefore

$$
\begin{aligned}
X_T &= 2\pi(f)(L_T) \\
&= 2\pi(f)(L_1 + L_2) \\
&= 2\pi(f)(L_1) + 2\pi(f)(L_2) \\
&= X_{L1} + X_{L2}
\end{aligned}
$$

Therefore, total series reactance is equal to the sum of the individual reactances, just as total series resistance is equal to the sum of the individual resistances.

Refer to the circuit of Figure 28-3.

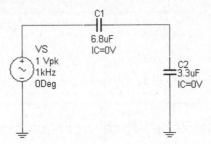

Figure 28-3: Example Series Capacitive Reactance Circuit

For each frequency in Table 28-3 calculate the individual capacitive reactances, the total capacitive reactance, and the total current for the circuit.

Table 28-3: Calculated Values for Example Series Capacitive Reactance Circuit

f	X_{C1}	X_{C2}	$X_T = X_{C1} + X_{C2}$	$I_T = VS / X_T$
10 Hz				
20 Hz				
50 Hz				
100 Hz				
200 Hz				
500 Hz				
1 kHz				
2 kHz				
5 kHz				
10 kHz				

Refer to the circuit of Figure 28-4.

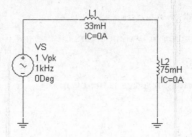

Figure 28-4: Example Series Inductive Reactance Circuit

For each frequency in Table 28-4 calculate the individual inductive reactances, the total inductive reactance, and the total current for the circuit.

Table 28-4: Calculated Values for Example Series Inductive Reactance Circuit

f	X_{L1}	X_{L2}	$X_T = X_{L1} + X_{L2}$	$I_T = VS / X_T$
10 Hz				
20 Hz				
50 Hz				
100 Hz				
200 Hz				
500 Hz				
1 kHz				
2 kHz				
5 kHz				
10 kHz				

28.2.3 Parallel Reactance

The equation for the total capacitance of two capacitors in parallel is

$$C_T = C_1 + C_2$$

The total reactance of two capacitors in parallel is therefore

$$X_T = 1 / [2\pi(f)(C_T)]$$

So

$$1 / X_T = 2\pi(f)(C_T)$$
$$= 2\pi(f)(C_1 + C_2)$$
$$= 2\pi(f)(C_1) + 2\pi(f)(C_2)$$
$$= 1 / X_{C1} + 1 / X_{C2}$$

Therefore

$$X_T = 1 / (1/X_{C1} + 1/X_{C2})$$

The equation for the total inductance of two inductors in parallel is

$$L_T = 1/(1/L_1 + 1/L_2) = L_1 L_2 / (L_1 + L_2)$$

The total reactance of the inductors in parallel is therefore

$X_T = 2\pi(f)(L_T)$

$\quad = 2\pi(f)[L_1 L_2 / (L_1 + L_2)]$

$\quad = [2\pi(f)(L_1 L_2)] / (L_1 + L_2)$

So

$1 / X_T = (L_1 + L_2) / [2\pi(f)(L_1 L_2)]$

$\quad = L_1 / [2\pi(f)(L_1 L_2)] + L_2 / [2\pi(f)(L_1 L_2)]$

$\quad = 1 / [2\pi(f)(L_2)] + 1 / [2\pi(f)(L_1)]$

$\quad = 1 / X_{L2} + 1 / X_{L2} = 1 / X_{L1} + 1 / X_{L2}$

Therefore

$X_T = 1/(1 / X_{L1} + 1 / X_{L2})$

Therefore, total parallel reactance is equal to the reciprocal of the sum of the reciprocals of the individual reactances, just as total parallel resistance is equal to the reciprocal of the sum of the reciprocals of the individual resistances.

Refer to the circuit of Figure 28-5.

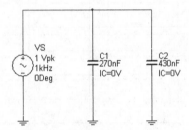

Figure 28-5: Example Parallel Capacitive Reactance Circuit

For each frequency in Table 28-5 calculate the individual capacitive reactances, the total capacitive reactance, and the total current for the circuit.

Table 28-5: Calculated Values for Example Parallel Capacitive Reactance Circuit

f	X_{C1}	X_{C2}	$X_T = 1/(1 / X_{C1} + 1 / X_{C2})$	$I_T = VS / X_T$
10 Hz				
20 Hz				
50 Hz				
100 Hz				
200 Hz				
500 Hz				
1 kHz				
2 kHz				
5 kHz				
10 kHz				

Refer to the circuit of Figure 28-6.

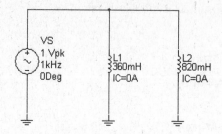

Figure 28-6: Example Parallel Inductive Reactance Circuit

For each frequency in Table 28-6 calculate the individual inductive reactances, the total inductive reactance, and the total current for the circuit.

Table 28-6: Calculated Values for Example Parallel Inductive Reactance Circuit

f	X_{L1}	X_{L2}	$X_T = 1/(1/X_{L1} + 1/X_{L2})$	$I_T = VS / X_T$
10 Hz				
20 Hz				
50 Hz				
100 Hz				
200 Hz				
500 Hz				
1 kHz				
2 kHz				
5 kHz				
10 kHz				

28.3 Design Verification

Once you calculate the reactances for capacitors or inductors in an ac circuit, you can use the basic concepts of series and parallel circuits and electronic laws such as Ohm's Law with which you analyzed dc resistive circuits. To verify your calculations you will now use an instrument with which you made dc measurements: the multimeter. The only difference is that you will use the ac measurement mode rather than dc measurement mode.

When you use the multimeter to make ac measurements, you must remember that the multimeter shows all values as rms (root-mean-square) values and not peak values. Use the following formulas to convert from peak to rms or vice versa.

RMS = Peak × 0.707

Peak = RMS × 1.414

The Multisim ac voltage source values in the pre-lab circuits are peak values, so your calculated current values are peak values as well. To verify the current values in the Multisim circuits you must calculate the ac ammeter (I_T (rms)) readings by 1.414 to determine the peak current value.

Open the circuit file Ex28-01. Set the voltage source to each of the frequencies in Table 28-7, use the multimeter to measure and record the rms total current in the circuit, and calculate the peak total current for each frequency.

Table 28-7: Measured Total Current for Ex28-01

f	I_T (rms)	I_T (peak)	f	I_T (rms)	I_T (peak)
10 Hz			500 Hz		
20 Hz			1 kHz		
50 Hz			2 kHz		
100 Hz			5 kHz		
200 Hz			10 kHz		

Do your peak total current values in Table 28-7 match those in Table 28-1?

Open the circuit file Ex28-02. Set the voltage source to each of the frequencies in Table 28-8, use the multimeter to measure and record the rms total current in the circuit, and calculate the peak total current for each frequency.

Table 28-8: Measured Total Current for Ex28-02

f	I_T (rms)	I_T (peak)	f	I_T (rms)	I_T (peak)
10 Hz			500 Hz		
20 Hz			1 kHz		
50 Hz			2 kHz		
100 Hz			5 kHz		
200 Hz			10 kHz		

Do your peak total current values in Table 28-8 match those in Table 28-2?

Open the circuit file Ex28-03. Set the voltage source to each of the frequencies in Table 28-9, use the multimeter to measure and record the rms total current in the circuit, and calculate the peak total current for each frequency.

Table 28-9: Measured Total Current for Ex28-03

f	I_T (rms)	I_T (peak)	f	I_T (rms)	I_T (peak)
10 Hz			500 Hz		
20 Hz			1 kHz		
50 Hz			2 kHz		
100 Hz			5 kHz		
200 Hz			10 kHz		

Do your peak total current values in Table 28-9 match those in Table 28-3?

Open the circuit file Ex28-04. Set the voltage source to each of the frequencies in Table 28-10, use the multimeter to measure and record the rms total current in the circuit, and calculate the peak total current for each frequency.

Table 28-10: Measured Total Current for Ex28-04

f	I_T (rms)	I_T (peak)	f	I_T (rms)	I_T (peak)
10 Hz			500 Hz		
20 Hz			1 kHz		
50 Hz			2 kHz		
100 Hz			5 kHz		
200 Hz			10 kHz		

Do your peak total current values in Table 28-10 match those in Table 28-4?

Open the circuit file Ex28-05. Set the voltage source to each of the frequencies in Table 28-11, use the multimeter to measure and record the rms total current in the circuit, and calculate the peak total current for each frequency.

Table 28-11: Measured Total Current for Ex28-05

f	I_T (rms)	I_T (peak)	f	I_T (rms)	I_T (peak)
10 Hz			500 Hz		
20 Hz			1 kHz		
50 Hz			2 kHz		
100 Hz			5 kHz		
200 Hz			10 kHz		

Do your peak total current values in Table 28-11 match those in Table 28-5?

Open the circuit file Ex28-06. Set the voltage source to each of the frequencies in Table 28-12, use the multimeter to measure and record the rms total current in the circuit, and calculate the peak total current for each frequency.

Table 28-12: Measured Total Current for Ex28-06

f	I_T (rms)	I_T (peak)	f	I_T (rms)	I_T (peak)
10 Hz			500 Hz		
20 Hz			1 kHz		
50 Hz			2 kHz		
100 Hz			5 kHz		
200 Hz			10 kHz		

Do your peak total current values in Table 28-12 match those in Table 28-6?

28.4 Application Exercise

Because you can calculate the total reactance for series or parallel reactive circuits the same way you would calculate the total resistance for series or parallel resistive circuits, it seems logical that you should be able to calculate the total reactance for series-parallel reactive circuits the same way you would calculate the total resistance for series-parallel resistive circuits. Refer to the circuit in Figure 28-7.

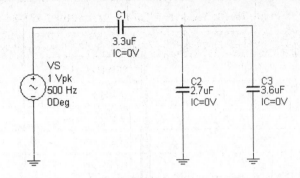

Figure 28-7: Series-Parallel Capacitive Reactance Circuit

If you replaced the capacitors with resistors the equation for the total circuit resistance would be

$$R_T = R1 + R2 \parallel R3$$

Therefore the equation for total circuit reactance should be

$$X_T = X_{C1} + X_{C2} \parallel X_{C2}$$

Calculate the values for each of the reactances and total current shown in Table 28-13 and record them in the "Calculated" column.

Table 28-13: Calculated and Measured Values for Series-Parallel Reactive Circuit

Circuit Value	Calculated	Measured
X_{C1}		
X_{C2}		
X_{C3}		
$X_{EQ} = X_{C2} \parallel X_{C3}$		
$X_T = X_{C1} + X_{EQ}$		
$I_T = V_S / X_T$		

Simulate the circuit file Pr28-01. Use the multimeter to measure the rms total current, calculate the peak total current, and record the peak value in the "Measured" column. Do the calculated and measured current values for the series-parallel reactive circuit agree?

28.5 Section Summary

This section introduced the concept of reactance and investigated the characteristics of capacitive and inductive reactance. Capacitive reactance decreases as frequency increases, and inductive reactance increases as frequency increases. You can work with reactance in ac circuits just as you would with resistance in dc circuits when calculating the total series, parallel, and series-parallel reactance for a circuit and can use Ohm's Law to calculate ac current from ac voltage and reactance.

29. Introduction to Complex Circuits

29.1 Introduction

In previous sections you worked with circuits that were either resistive or reactive. A complex circuit is a circuit that contains both resistive and reactive components. The term "complex" does not refer to how complicated the circuit is. It refers instead to the electrical quantities that are associated with the circuit and that consist of both a real and an imaginary component. The real component represents the resistive aspect of the circuit for which the voltage and current are in phase. The imaginary component represents the reactive aspect of the circuit for which the voltage and current are 90° out of phase.

A study of complex circuit requires special objects that can represent both magnitude and phase. These objects are called **phasors**. Refer to Figure 29-1, which shows a complex series circuit and the phasor representation of its components.

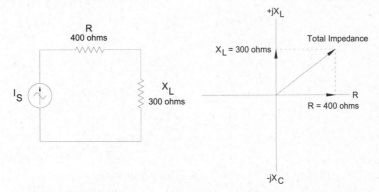

Figure 29-1: Complex Series Circuit and Phasor Representation

As Figure 29-1 shows, the resistance is a 400 Ω phasor along the positive real axis labeled "R", and the inductive reactance is a 300 Ω phasor along the positive imaginary axis. The total impedance, or combined resistance and reactance, is a phasor that lies somewhere between the positive real and positive imaginary axes. Phasor arithmetic allows you to determine the magnitude and phase for the total series impedance.

In this section you will:

- Review the basics of phasor arithmetic.

- Use the Multisim program to verify the basics of phasor arithmetic for complex circuits.

29.2 Pre-Lab

You can represent phasors in either rectangular or polar form.

Rectangular form expresses the phasor as a sum of the real and imaginary components and uses the prefix **j** to signify the imaginary component. As an example, the rectangular form expresses the impedance Z_T for a circuit with resistance R and inductive reactance X as

$$Z_T = R + jX$$

The advantage of rectangular form is that you can immediately determine the magnitudes of the real and imaginary portions of the complex value. This simplifies addition and subtraction of phasors, such as when you are calculating series impedances or working with Kirchhoff's Voltage and Current Laws. For the impedances $Z_1 = R_1 + jX_1$ and $Z_2 = R_2 + jX_2$ then

$$Z_3 = Z_1 + Z_2 = (R_1 + jX_1) + (R_2 + jX_2) = (R_1 + R_2) + j(X_1 + X_2)$$

$$Z_4 = Z_1 - Z_2 = (R_1 + jX_1) - (R_2 + jX_2) = (R_1 - R_2) + j(X_1 - X_2)$$

Polar form shows the magnitude and phase of the phasor. For a circuit with impedance Z and a phase angle of φ (phi) polar form represents the phasor for the impedance Z_T as

$$Z_T = Z \angle \varphi$$

The advantage of polar form is that you can immediately determine the magnitude and phase angle of the complex value. This simplifies multiplication and division of phasors, such as when you working with Ohm's Law. If a voltage $V_S = V \angle \varphi$ and current $I_T = I \angle \theta$ then the power is

$$P_T = V_S \times I_T = (V \angle \varphi) \times (I \angle \theta) = (V \times I) \angle (\varphi + \theta)$$

Similarly

$$X_T = V_S / I_T = (V \angle \varphi) / (I \angle \theta) = (V / I) \angle (\varphi - \theta)$$

If your calculator cannot perform complex arithmetic directly and does not have rectangular to polar conversion functions, you can use conversion equations to select the phasor form that is most convenient for your calculations. Refer to Figure 29-2.

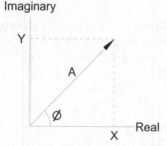

Figure 29-2: Phasor Rectangular and Polar Forms

To convert rectangular form $X + jY$ to polar form $A \angle \varphi$:

Amplitude: $A = \sqrt{X^2 + Y^2}$ (from Pythagorean Theorem)

Phase: $\varphi = \tan^{-1}(Y / X)$ (from standard trigonometry)

To convert polar form $A \angle \varphi$ to rectangular form $X + jY$:

Real component: $X = A \cos(\varphi)$ (from standard trigonometry)

Imaginary component: $Y = A \sin(\varphi)$ (from standard trigonometry)

Refer to the circuit in Figure 29-3.

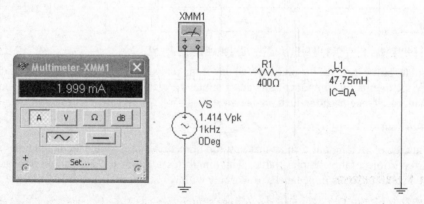

Figure 29-3: Example Complex Circuit

The frequency of the voltage source is 1 kHz so $X_{L1} = 2\pi(f)(L_1) = 2\pi(1 \text{ kHz})(47.75 \text{ mH}) = 300 \text{ }\Omega$. The total impedance is therefore $Z_T = R1 + jX_{L1} = 400 \text{ }\Omega + j300 \text{ }\Omega$.

From Ohm's Law, $I_T = V_S / Z_T$. Polar form is more convenient than rectangular form for multiplication and division problems. Converting $R1 + jX_{L1}$ to polar form $Z \angle \varphi$ gives

Amplitude: $Z = \sqrt{R1^2 + X_{L1}{}^2} = \sqrt{(400 \text{ }\Omega)^2 + (300 \text{ }\Omega)^2} = 500 \text{ }\Omega$

Phase: $\varphi = \tan^{-1}(X_{L1} / R1) = \tan^{-1}(300 \text{ }\Omega / 400 \text{ }\Omega) = \tan^{-1}(0.75) = 36.9°$

so $Z_T = 500 \text{ }\Omega \angle 36.9°$. Therefore

$I_T = V_S / Z_T$

$\quad = (1.414 \text{ Vpk} \angle 0°) / (500° \angle 36.9°)$

$\quad = (1.414 \text{ Vpk} / 500 \text{ }\Omega) \angle (0° - 36.9°)$

$\quad = 2.828 \text{ mApk} \angle -36.9°$

The ac current amplitude is 2.828 mApk. Note that this is the **peak** current, as the source voltage value is in units of peak volts (Vpk). However, the multimeter's ac ammeter mode measures values in rms. The rms amplitude is 0.707 times the peak value, so the rms current value is

$I_T = (0.707)(2.828 \text{ mApk})$

$\quad = 2.000 \text{ mA}_{\text{RMS}}.$

This matches the multimeter measurement shown in Figure 29-3.

Calculate the rms amplitude of I_T for each of the frequencies shown in Table 29-1.

Table 29-1: Calculated RMS Current Amplitudes for Example Complex Circuit

f	$X_{L1} = 2\pi(f)(L1)$	$Z_T = R1 + jX_{L1}$	$Z = \sqrt{R1^2 + X_{L1}{}^2}$	$I_T(\text{pk}) = V_S / Z$	$I_T(\text{rms}) =$ $0.707 \times I_T(\text{pk})$
0					
1 kHz					
2 kHz					
3 kHz					
4 kHz					
5 kHz					
6 kHz					
7 kHz					
8 kHz					
9 kHz					
10 kHz					

29.3 Design Verification

Open the circuit file Ex29-01. Set the voltage source to each of the frequencies shown in Table 29-2, simulate the circuit file, and record the measured rms current on the multimeter for each frequency.

Table 29-2: Measured RMS Current Amplitudes for Example Complex Circuit

f	I_T(rms)	f	I_T(rms)
1 kHz		6 kHz	
2 kHz		7 kHz	
3 kHz		8 kHz	
4 kHz		9 kHz	
5 kHz		10 kHz	

Do the calculated I_T(rms) values in Table 29-1 and measured I_T(rms) values in Table 29-2 agree?

29.4 Application Exercise

Refer to the circuit in Figure 29-4.

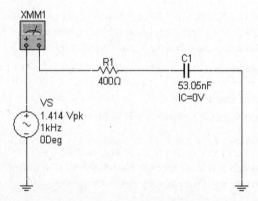

Figure 29-4: Practice Complex Circuit

Calculate the rms amplitude of I_T for each of the frequencies shown in Table 29-3.

Table 29-3: Calculated RMS Current Amplitudes for Example Complex Circuit

f	$X_{C1} = 1/[2\pi(f)(C1)]$	$Z_T = R1 - jX_{C1}$	$Z = \sqrt{R1^2 + X_{C1}^2}$	I_T(pk) $= V_S / Z$	I_T(rms) $= 0.707 \times I_T$(pk)
1 kHz					
2 kHz					
3 kHz					
4 kHz					
5 kHz					
6 kHz					
7 kHz					
8 kHz					
9 kHz					
10 kHz					

Open the circuit file Pr29-01. Set the voltage source to each of the frequencies shown in Table 29-4, simulate the circuit file, and record the measured rms current on the multimeter for each frequency.

Table 29-4: Measured RMS Current Amplitudes for Example Complex Circuit

f	I_T(rms)	f	I_T(rms)
1 kHz		6 kHz	
2 kHz		7 kHz	
3 kHz		8 kHz	
4 kHz		9 kHz	
5 kHz		10 kHz	

Do the calculated I_T(rms) values in Table 29-3 and measured I_T(rms) values in Table 29-4 agree?

29.5 Section Summary

This section introduced the concept of complex circuits, which consist of both resistive and reactive components. The opposition to current flow in complex circuits is impedance, which has both a resistive (real) and reactive (imaginary) component. Complex circuit analysis requires the use of phasor arithmetic to account for both the magnitude and phase of the electrical quantities. Ac multimeters can measure only the magnitude, or amplitude, of complex quantities, but can still verify basic phasor concepts in complex circuits.

Ac circuit analysis often requires you to add, subtract, multiply, and divide phasor quantities for the same circuit. Because of this, you should convert the phasor values you are using into both rectangular and polar forms so that you can readily use them in subsequent calculations.

30. AC Response of RC Circuits

30.1 Introduction

The previous section introduced the concept of complex circuits, which contain both resistive and reactive elements. Electrical quantities in complex circuits have both magnitude and phase. You can measure magnitude with basic ac meters, but phase measurements require the use of an oscilloscope.

RC circuits include series, parallel, and series-parallel configurations. The ac analysis of RC circuits is similar to resistor circuit analysis except that calculations must use complex arithmetic.

In this section you will:

- Study the techniques for analyzing series, parallel, and series-parallel RC circuits.

- Use the Multisim software to verify your ac analysis results for RC circuits.

30.2 Pre-Lab

30.2.1 Series RC Circuits

The basic procedure for analyzing series RC circuits is

1) Calculate the capacitive reactance for all capacitors.

2) Combine all resistive and reactive terms.

3) Convert the total series impedance to polar form.

4) Use V_S and Z_T to calculate the total series current.

5) Use the total series current and component values to calculate the component voltages.

Refer to the series RC circuit and its associated phasors in Figure 30-1.

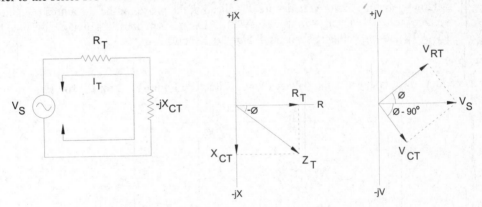

Figure 30-1: Series RC Circuit and Associated Phasors

Step 1: Calculate the capacitive reactance for all capacitors. The capacitive reactance is equal to

$$X_C = 0 - j(1 / 2\pi fC) = (1 / 2\pi fC) \angle -90°$$

Step 2: Combine all resistive and reactive terms. For a series RC circuit that consists of M resistors and N capacitors the total impedance is

$$Z_T = (R_1 + R_2 + ... + R_M) - j(X_{C1} + X_{C2} + ... + X_{CN})$$

$$= R_T - jX_{CT}$$

Step 3: **Convert the total series impedance to polar form.** For the rectangular form $Z_T = R_T - jX_{CT} = Z_T$ $\angle \varphi$, the amplitude and phase angle are

$$\text{Amplitude:} \quad Z_T = \sqrt{R_T{}^2 + X_{CT}{}^2}$$

$$\text{Phase:} \quad -\varphi = \tan^{-1}(-X_{CT} / R_T) = -\tan^{-1}(X_{CT} / R_T)$$

For a series RC circuit the phase angle for the impedance is always a negative value between $0°$ and $-90°$. To avoid confusion this discussion will use φ to express a positive angle and $-\varphi$ to express a negative angle.

Step 4: **Use V_S and Z_T to calculate the total series current.** The voltage does not specify a phase, so assume the voltage phase is $0°$. Then

$$I_T = (V_S \angle 0) / (Z_T \angle -\varphi)$$

$$= (V_S / Z_T) \angle \varphi$$

Note that the current phase is the opposite of the impedance phase and is a positive value. The current therefore leads the voltage by φ.

Step 5: **Use the total series current and component values to calculate the component voltages.** For each resistor

$$V_R = (R \angle 0)(I_T \angle \varphi)$$

$$= (R \times I_T) \angle \varphi$$

For each capacitor

$$V_C = (X_C \angle -90°)(I_T \angle \varphi)$$

$$= (X_C \times I_T) \angle (-90° + \varphi) = (X_C \times I_T) \angle (\varphi - 90°)$$

Note that the resistor voltage leads the source voltage by φ and that the capacitor voltage is $90°$ behind the phase for the resistors (i.e., the capacitor voltages lag the resistor voltages by $90°$). Alternatively the resistor voltages are $90°$ ahead of the phase for the capacitors (i.e., the resistor voltages lead the capacitor voltages by $90°$). Also, per Kirchhoff's Voltage Law, note that $V_S = V_{RT} + V_{CT}$, as the voltage phasors in Figure 30-1 show.

Example:

Determine the total current and voltages for the circuit of Figure 30-2 for $f = 100$ Hz.

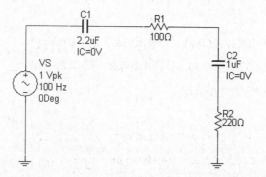

Figure 30-2: Example Series RC Circuit

Step 1: **Calculate the capacitive reactance for all capacitors.** For $C1 = 2.2$ µF and $C2 = 1$ µF

$$X_{C1} = -j[1 / (2\pi)(100 \text{ Hz})(2.2 \text{ µF})] = -j(723 \text{ } \Omega)$$

$$X_{C2} = -j[1 / (2\pi)(100\ \text{Hz})(1\ \mu\text{F})] = -j(1.59\ \text{k}\Omega)$$

Step 2: **Combine all resistive and reactive terms.**

$$\begin{aligned} Z_T &= -j(723\ \Omega) + 100\ \Omega - j(1.59\ \text{k}\Omega) + 220\ \Omega \\ &= (100\ \Omega + 220\ \Omega) - j(723\ \Omega + 1.59\ \text{k}\Omega) \\ &= 320\ \Omega - j(2.31\ \text{k}\Omega) \end{aligned}$$

Step 3: **Convert the total series impedance to polar form.** For $Z_T = 320\ \Omega - j(2.31\ \text{k}\Omega)$

Amplitude: $Z_T = \sqrt{(320\ \Omega)^2 + (2.31\ \text{k}\Omega)^2} = 2.34\ \text{k}\Omega$

Phase: $\varphi = -\tan^{-1}(2.31\ \text{k}\Omega / 320\ \Omega) = -\tan^{-1}(7.23) = -82.1°$

Polar form: $Z_T = 2.34\ \text{k}\Omega \angle -82.1°$

Step 4: **Use V_S and Z_T to calculate the total current.** $V_S = 1\ \text{Vpk} \angle 0°$ so

$$\begin{aligned} I_T &= V_S / Z_T \\ &= (1\ \text{Vpk} \angle 0°) / (2.34\ \text{k}\Omega \angle -82.1°) \\ &= 428\ \mu\text{Apk} \angle 82.1° \end{aligned}$$

Step 5: **Use the total series current and component values to calculate the component voltages.** For each of the components the voltages in polar and rectangular form are

$$\begin{aligned} V_{C1} = I_T \times X_{C1} &= (428\ \mu\text{Apk} \angle 82.1°)(723\ \Omega \angle -90°) \\ &= 310\ \text{mVpk} \angle -7.9° = 307\ \text{mVpk} - j(42.6\ \text{mVpk}) \end{aligned}$$

$$\begin{aligned} V_{R1} = I_T \times R1 &= (428\ \mu\text{Apk} \angle 82.1°)(100\ \Omega \angle 0°) \\ &= 42.8\ \text{mVpk} \angle 82.1° = 5.9\ \text{mVpk} + j(42.4\ \text{mVpk}) \end{aligned}$$

$$\begin{aligned} V_{C2} = I_T \times X_{C2} &= (428\ \mu\text{Apk} \angle 82.1°)(1.59\ \text{k}\Omega \angle -90°) \\ &\quad 681\ \text{mVpk} \angle -7.9° = 675\ \text{mVpk} - j(93.5\ \text{mVpk}) \end{aligned}$$

$$\begin{aligned} V_{R2} = I_T \times R2 &= (428\ \mu\text{Apk} \angle 82.1°)(220\ \Omega \angle 0°) \\ &= 94.1\ \text{mVpk} \angle 82.1° = 12.9\ \text{mVpk} + j(93.2\ \text{mVpk}) \end{aligned}$$

As a check,

$$\begin{aligned} V_{RT} = V_{R1} + V_{R2} \\ &= 5.9\ \text{mVpk} + j(42.4\ \text{mVpk}) + 12.9\ \text{mVpk} + j(93.2\ \text{mVpk}) \\ &= (5.9\ \text{mVpk} + 12.9\ \text{mVpk}) + j(42.4\ \text{mVpk} + 93.2\ \text{mVpk}) \\ &= 18.8\ \text{mVpk} + j(136\ \text{mVpk}) \end{aligned}$$

$$\begin{aligned} V_{CT} = V_{C1} + V_{C2} \\ &= 307\ \text{mVpk} - j(42.6\ \text{mVpk}) + 675\ \text{mVpk} - j(93.5\ \text{mVpk}) \\ &= (307\ \text{mVpk} + 675\ \text{mVpk}) - j(42.6\ \text{mVpk} + 93.5\ \text{mVpk}) \\ &= 982\ \text{mVpk} - j(136\ \text{mVpk}) \end{aligned}$$

So

$$\begin{aligned} V_S &= V_{RT} + V_{CT} \\ &= 18.8\ \text{mVpk} + j(136\ \text{mVpk}) + 982\ \text{mVpk} - j(136\ \text{mVpk}) \end{aligned}$$

$$= (18.8 \text{ mVpk} + 982 \text{ mVpk}) + j(136 \text{ mVpk} - 136 \text{ mVpk})$$

$$= 1.001 \text{ Vpk} - j(0 \text{ mVpk}) \approx 1.0 \text{ Vpk} + j(0.0 \text{ Vpk}) = 1.0 \text{ Vpk} \angle 0°$$

The voltages comply with Kirchhoff's Voltage Law. The minor deviation is due to rounding in calculations.

Now, calculate the voltages and currents for the circuit of Figure 30-2 for $f = 1$ kHz and $f = 10$ kHz in both polar and rectangular form and record the values in Table 30-1.

Table 30-1: Calculated Values for Example Series RC Circuit

Circuit Value	$f = 1$ kHz		$f = 10$ kHz	
	Rectangular	Polar	Rectangular	Polar
$X_{C1} = 1 / [2\pi(f)(C1)]$				
$X_{C2} = 1 / [2\pi(f)(C2)]$				
$R_T = R1 + R2$				
$X_{CT} = X_{C1} + X_{C2}$				
$Z_T = R_T + X_{CT}$				
$I_T = V_S / Z_T$				
$V_{C1} = I_T \times X_{C1}$				
$V_{C2} = I_T \times X_{C2}$				
$V_{R1} = I_T \times R1$				
$V_{R2} = I_T \times R2$				

30.2.2 Parallel RC Circuits

The basic procedure for analyzing parallel RC circuits is

1) Calculate the capacitive reactance for all capacitors.

2) Use V_S and the branch resistances and reactances to calculate the branch currents.

3) Add the branch currents to determine the total current.

4) Use V_S and I_T to calculate the total impedance.

Refer to the parallel RC circuit and associated current phasors in Figure 30-3.

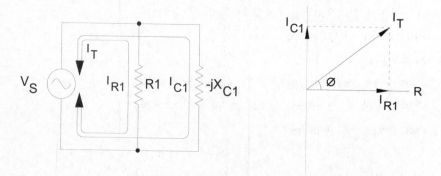

Figure 30-3: Parallel RC Circuit and Associated Phasors

Step 1: Calculate the capacitive reactance for all capacitors. The capacitive reactance is equal to

$$X_C = 0 - j(1 / 2\pi fC) = (1 / 2\pi fC) \angle -90°$$

Step 2: Use V_S and the branch resistances and reactances to calculate the branch currents. The voltage does not specify a phase so assume the voltage phase is 0°. The current for each resistive branch is

$$I_R = (V_S \angle 0°) / (R \angle 0°)$$
$$= (V_S / R) \angle 0° = (V_S / R) + j(0 \text{ A})$$

The current for each reactive branch is

$$I_C = (V_S \angle 0°) / (X_C \angle -90°)$$
$$= (V_S / X_C) \angle 90° = 0 \text{ A} + j(V_S / X_C)$$

Step 3: Add the branch currents to determine the total current. From Step 2 this is

$$I_T = (V_S / R_1) + ... + (V_S / R_M) + j[(V_S / X_{C1}) + ... + (V_S / X_{CN})]$$
$$= V_S(1/R_1 + ... + 1/R_M) + j[V_S(1/X_{C1} + ... + 1/X_{CN})]$$

If we let $R_{EQ} = 1 / (1/R_1 + ... + 1/R_M)$ and $X_{EQ} = 1/ (1/X_{C1} + ... + 1/X_{CN})$ this becomes

$$I_T = V_S / R_{EQ} + j(V_S / X_{EQ})$$
$$= I_{REQ} + j(I_{XEQ})$$

Amplitude: $I_T = \sqrt{(I_{REQ})^2 + (I_{XEQ})^2}$

Phase: $\varphi = \tan^{-1}(I_{XEQ} / I_{REQ})$

For a parallel RC circuit the phase angle for the current is always a positive value between 0° and +90°.

Step 4: Use V_S and I_T to calculate the total impedance. Then

$$Z_T = (V_S \angle 0) / (I_T \angle \varphi)$$
$$= V_S / I_T \angle (0 - \varphi)$$
$$= (V_S / I_T) \angle -\varphi$$

Note that the impedance phase is the opposite of the current phase and is negative.

Example:

Determine the currents and total impedance for the circuit of Figure 30-4 for f = 100 Hz.

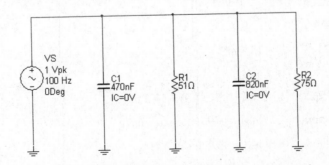

Figure 30-4: Example Parallel RC Circuit

Step 1: Calculate the capacitive reactance for all capacitors. For $C1$ = 470 nF and $C2$ = 820 nF

$$X_{C1} = -j[1 / (2\pi)(100 \text{ Hz})(470 \text{ nF})] = -j(3.39 \text{ k}\Omega) = 3.39 \text{ k}\Omega \angle -90°$$

$$X_{C2} = -j[1 / (2\pi)(100 \text{ Hz})(820 \text{ nF})] = -j(1.94 \text{ k}\Omega) = 1.94 \text{ k}\Omega \angle -90°$$

Step 2: Use V_S and the branch resistances and reactances to calculate the branch currents.

$$I_{C1} = (1.0 \text{ Vpk} \angle 0°) / (3.39 \text{ k}\Omega \angle -90°)$$

$$= 295 \text{ }\mu\text{Apk} \angle 90° = j(295 \text{ }\mu\text{Apk})$$

$$I_{R1} = (1.0 \text{ Vpk} \angle 0°) / (51 \text{ }\Omega \angle 0°)$$

$$= 19.6 \text{ mA} \angle 0° = 19.6 \text{ mApk}$$

$$I_{C2} = (1.0 \text{ Vpk} \angle 0°) / (1.94 \text{ k}\Omega \angle -90°)$$

$$= 515 \text{ }\mu\text{Apk} \angle 90° = j(515 \text{ }\mu\text{Apk})$$

$$I_{R2} = (1.0 \text{ Vpk} \angle 0°) / (75 \text{ }\Omega \angle 0°)$$

$$= 13.3 \text{ mA} \angle 0° = 13.3 \text{ mApk}$$

Step 3: Add the branch currents to determine the total current. The rectangular and polar forms are

$$I_T = j(295 \text{ }\mu\text{Apk}) + 19.6 \text{ mApk} + j(515 \text{ }\mu\text{Apk}) + 13.3 \text{ mApk}$$

$$= (19.6 \text{ mApk} + 13.3 \text{ mApk}) + j(295 \text{ }\mu\text{Apk} + 515 \text{ }\mu\text{Apk})$$

$$= 32.9 \text{ mApk} + j(810 \text{ }\mu\text{Apk}) = 32.9 \text{ mA} \angle 1.41°$$

Step 4: Use V_S and I_T to calculate the total impedance. $V_S = 1 \text{ Vpk} \angle 0°$ so

$$Z_T = V_S / I_T$$

$$= (1 \text{ Vpk} \angle 0°) / (32.9 \text{ mApk} \angle 1.41°)$$

$$= 30.4 \text{ }\Omega \angle -1.41°$$

As a check, determine Z_T by calculating the reciprocal of the sum of the reciprocals for each resistance and reactance. Note that the unit for the reciprocal of the ohm is the Siemen (S).

$$1 / X_{C1} = (1 \angle 0°) / (3.39 \text{ k}\Omega \angle -90°)$$

$$= 295 \text{ }\mu\text{S} \angle 90° = j(295 \text{ }\mu\text{S})$$

$$1 / R1 = (1 \angle 0°) / (51 \text{ }\Omega \angle 0°)$$

$$= 19.6 \text{ mS} \angle 0° = 19.6 \text{ mS}$$

$$1 / X_{C2} = (1 \angle 0°) / (1.94 \text{ k}\Omega \angle -90°)$$

$$= 515 \text{ }\mu\text{S} \angle 90° = j(515 \text{ }\mu\text{S})$$

$$1 / R2 = (1 \angle 0°) / (75 \text{ }\Omega \angle 0°)$$

$$= 13.3 \text{ mS} \angle 0° = 13.3 \text{ mS}$$

From this

$$1 / Z_T = j(295 \text{ }\mu\text{S}) + 19.6 \text{ mS} + j(515 \text{ }\mu\text{S}) + 13.3 \text{ mS}$$

$$= (19.6 \text{ mS} + 13.3 \text{ mS}) + j(295 \text{ mS} + 515 \text{ mS})$$

$$= 32.9 \text{ mS} + j(810 \text{ }\mu\text{S}) = 32.9 \text{ mS} \angle 1.41°$$

Therefore

$$Z_T \quad = (1 \angle 0°) / (32.9 \text{ mS} \angle 1.41°)$$

$$= 30.4 \ \Omega \angle -1.41°$$

The total impedance values check.

Now, calculate the currents and total impedance for the circuit of Figure 30-4 for f = 1 kHz and f = 10 kHz in both polar and rectangular form and record the values in Table 30-2.

Table 30-2: Calculated Values for Example Parallel RC Circuit

Circuit Value	f = 1 kHz		f = 10 kHz	
	Rectangular	Polar	Rectangular	Polar
$X_{C1} = 1 / [2\pi(f)(C1)]$				
$X_{C2} = 1 / [2\pi(f)(C2)]$				
$I_{R1} = V_S / R1$				
$I_{R2} = V_S / R2$				
$I_{C1} = V_S / X_{C1}$				
$I_{C2} = V_S / X_{C2}$				
$I_T = I_{R1} + I_{R2} + I_{C1} + I_{C2}$				
$Z_T = V_S / I_T$				

30.3 Design Verification

30.3.1 Phase Measurements

To measure phase between signals you must measure both signals at the same time. This is potentially a problem with a two-channel oscilloscope if one of the signals is across a floating component, as making a differential measurement across a floating component in itself requires both channels. One solution to this problem is to use supplies with isolated outputs so that there is no chance of shorting out part of the circuit through an oscilloscope probe reference. A second solution is to use an isolation transformer so that the circuit ground is not the same as the oscilloscope reference ground, but this can create a safety hazard.

In Multisim circuits the circuit ground is isolated from the oscilloscope reference ground, so that you can connect the + and – leads of the oscilloscope probes across floating components with no problem. Refer to Figure 30-5, which shows the connection for measuring the phase between VS and V_{R1}.

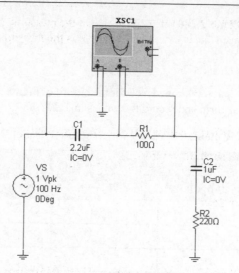

Figure 30-5: Example Phase Measurement Setup

In Figure 30-5 Channel A displays the waveform for *VS* and Channel 2 displays the waveform for V_{R1}.

Phase measurements require the following:

1) The frequency (and period) of both signals must be identical.

2) The measurements must be at corresponding points of the measured waveforms (positive zero-crossings, positive peaks, etc.).

3) The measurement points must be within 1/2 period of each other.

Refer to Figure 30-6.

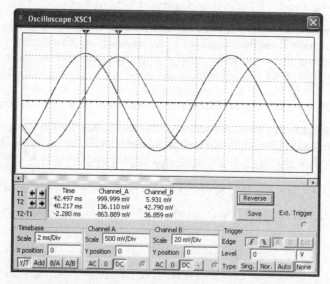

Figure 30-6: Example Phase Measurement

Although phase measurements are typically at the positive zero-crossing of both signals, they can also be at the positive maximum as Figure 30-6 shows. The advantage of using the positive peaks is that you can readily verify the amplitude of the signals as well as the phase between them. In the example phase measurement, Cursor 1 is measuring *VS* on Channel A and shows an amplitude of about 1.0 Vpk. Cursor 2 is measuring V_{R1} on Channel B and shows an amplitude of 42.790 mVpk. The time difference between the measured waveforms $T2 - T1$ is −2.280 ms, indicating that the measured point on V_{R1} happens before (leads) that of VS by 2.280 ms. The equation below converts the measured time difference into phase:

$$\varphi = (\Delta t\, /\, T) \times 360°$$

where

φ is the phase,

Δt is the measured time difference between the waveforms, and

T is the period of the measured waveforms.

For Figure 30-6 this gives

$$\varphi = (2.280 \text{ ms} / 10 \text{ ms}) \times 360°$$

$$= 0.228 \times 360°$$

$$= 82.1°$$

This gives $V_{R1} = 42.79$ mVpk $\angle$ 82.1° = 42.8 mVpk $\angle$ 82.1°. This agrees with the calculated V_{R1} value in the pre-lab for the series RC circuit.

30.3.2 Series RC Circuit

1. Open the circuit Ex30-01.

2. Select the two-channel oscilloscope from the Instrument toolbar.

3. Connect the "+" input of Channel A between VS and $C1$ and the "−" input to circuit ground.

4. Connect the inputs of Channel B across $C1$ with the "+" input closer to the positive terminal of VS.

5. Change the color of the "+" input of Channel B to a different color from the rest of the circuit wiring. This is not necessary but will help you to distinguish between the waveforms.

6. Click the **Run** switch to simulate the circuit, run the simulation for at least 50 ms, and click the **Run** switch to stop the simulation.

7. Adjust the "Scale" values of both channels so that each waveform is between four and six divisions high.

8. Adjust the timebase scale division so that no more than one cycle of each waveform is visible.

9. Adjust the scroll bar below the oscilloscope display so that you can see one positive peak for both waveforms. Only one positive peak of each should be visible.

10. Right-click Cursor 1 and select "Select Trace ID" from the cursor right-click menu.

11. Select "Channel A" from the drop-down list (if it is not already selected) and click the **OK** button.

12. Right-click Cursor 1 and select "Go to next Y_MAX =>" from the cursor right-click menu. The cursor should move to the positive peak of Channel A.

13. Right-click Cursor 2 and select "Select Trace ID" from the cursor right-click menu.

14. Select "Channel B" from the drop-down list (if it is not already selected) and click the **OK** button.

15. Right-click Cursor 2 and select "Go to next Y_MAX <=" from the cursor right-click menu. The cursor should move to the positive peak of Channel B.

16. Record the Channel B amplitude in Table 30-3 as the value of V_{C1} for $f = 100$ Hz.

17. Use the value of $T2 - T1$ to calculate the phase. Record this phase in Table 30-3 as the value of φ_{C1} for the "$f = 100$ Hz" column.

18. Repeat Steps 4 through 17 for *R1*, *C2*, and *R2* and record the amplitude and phase values in Table 30-3.

19. Change the frequency of the ac voltage source to 1 kHz and 10 kHz and repeat Steps 6 through 18.

Table 30-3: Measured Values for Example Series RC Circuit

Circuit Value	f = 100 Hz	f = 1 kHz	f = 10 kHz
V_{C1}			
φ_{C1}			
V_{R1}			
φ_{R1}			
V_{C2}			
φ_{C2}			
V_{R2}			
φ_{R2}			

Do your measured values in Table 30-3 agree with your calculated values in Table 30-1?

30.3.3 Parallel RC Circuit

You cannot directly measure currents for the example parallel RC circuit in Figure 30-4. Because the currents and voltages for a resistor are in phase, however, you can measure the voltage across a small sense resistor that is in series with each parallel branch and use the measured voltage to calculate the current. However, you must ensure that the sense resistance is no more than 1/100th the impedance of the branch. For this circuit $R = 1$ mΩ is acceptable. Note that *R1* and *R2* are the sense resistors for their branches.

1. Open the circuit Ex30-02.

2. Select the two-channel oscilloscope from the Instrument toolbar.

3. Connect the "+" input of Channel A between *VS* and *C1* and the "–" input to circuit ground.

4. Connect the "+" input of Channel B between *C1* and *RS1* and the "–" input to circuit ground.

5. Change the color of the "+" input of Channel B to a different color from the rest of the circuit wiring. This is not necessary but will help you to distinguish between the waveforms.

6. Click the **Run** switch to simulate the circuit, run the simulation to run for at least 50 ms, and click the **Run** switch to stop the simulation.

7. Adjust the "Scale" values of both channels so that each waveform is between four and six divisions high.

8. Adjust the timebase scale division so that no more than one cycle of each waveform is visible.

9. Adjust the scroll bar below the oscilloscope display so that you can see one positive peak for both waveforms. Only one positive peak of each should be visible.

10. Right-click Cursor 1 and select "Select Trace ID" from the cursor right-click menu.

11. Select "Channel A" from the drop-down list (if it is not already selected) and click the **OK** button.

12. Right-click Cursor 1 and select "Go to next Y_MAX =>" from the cursor right-click menu. The cursor should move to the positive peak of Channel A.

13. Right-click Cursor 2 and select "Select Trace ID" from the cursor right-click menu.

14. Select "Channel B" from the drop-down list (if it is not already selected) and click the **OK** button.

15. Right-click Cursor 2 and select "Go to next Y_MAX <=" from the cursor right-click menu. The cursor should move to the positive peak of Channel B.

16. Record the Channel B amplitude divided by the sense resistance in Table 30-4 as the value of I_{C1} for $f = 100$ Hz.

17. Use the value of $T2 - T1$ to calculate the phase. Record this phase in Table 30-4 as the value of φ_{C1} for the "$f = 100$ Hz" column.

18. Repeat Steps 4 through 17 for $R1$, $RS2$ and $C2$, and $R2$ and record the amplitude and phase values in Table 30-4.

19. Change the frequency of the ac voltage source to 1 kHz and 10 kHz and repeat Steps 4 through 18.

Table 30-4: Measured Values for Example Parallel RC Circuit

Circuit Value	$f = 100$ Hz	$f = 1$ kHz	$f = 10$ kHz
I_{C1}			
φ_{C1}			
I_{R1}			
φ_{R1}			
I_{C2}			
φ_{C2}			
I_{R2}			
φ_{R2}			

Do your measured values in Table 30-4 agree with your calculated values in Table 30-2?

30.4 Application Exercise

The ac analysis of series-parallel RC circuits is similar to the dc analysis of series-parallel resistive circuits, and combines the ac analysis techniques for series RC circuits and parallel RC circuits. You replace RC combinations in the circuit with complex circuit equivalents until you determine the total circuit impedance, and then work backwards to find the voltages and currents for individual circuit components.

Refer to the series-parallel RC circuit in Figure 30-7.

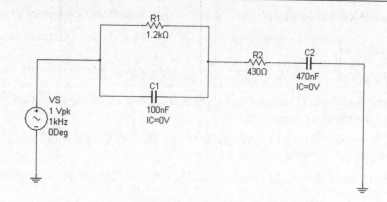

Figure 30-7: Example Series-Parallel RC Circuit

Calculate each of the values shown in Table 30-5 for the example series-parallel RC circuit.

Table 30-5: Calculated Values for Series-Parallel RC Circuit

Circuit Value	Rectangular Form	Polar Form
$X_{C1} = 1 / [2\pi(1 \text{ kHz})(100 \text{ nF})]$		
$X_{C2} = 1 / [2\pi(1 \text{ kHz})(470 \text{ nF})]$		
$Z_{EQ} = R1 \parallel X_{C1}$		
$Z_T = Z_{EQ} + R2 + X_{C2}$		
$I_T = I_{R2} = I_{C2} = V_S / Z_T$		
$V_{R1} = V_{C1} = I_T \times Z_{EQ}$		
$V_{R2} = I_T \times R2$		
$V_{C2} = I_T \times X_{C2}$		

Open the circuit file Pr30-01. Determine the amplitude and phase for each of the values shown in Table 30-6.

Table 30-6: Measured Values for Example Series-Parallel RC Circuit

Circuit Value	Measured (Polar Form)
I_T	
V_{C1}	
I_{C1}	
V_{R1}	
I_{R1}	
V_{C2}	
I_{C2}	
V_{R2}	
I_{R2}	

Do your measured values in Table 30-6 match your calculated values in Table 30-5?

30.5 Section Summary

This section covered the ac analysis of series, parallel, and series-parallel RC circuits. Ac analysis requires an oscilloscope to determine both the amplitude and phase components of each circuit element. This limits direct circuit measurement to ac voltages, but circuit and sense resistors allow you to determine the amplitude and phase of circuit currents.

31. AC Response of RL Circuits

31.1 Introduction

RL circuits include series, parallel, and series-parallel configurations. The ac analysis of RL circuits is very similar to that of RC circuits, except that the phase of inductive reactance is opposite that of capacitive reactance.

In this section you will:

- Study the techniques for analyzing series, parallel, and series-parallel RL circuits.

- Use the Multisim software to verify your ac analysis results for RL circuits.

31.2 Pre-Lab

31.2.1 Series RC Circuits

The basic procedure for analyzing series RL circuits is

1) Calculate the inductive reactance for all inductors.

2) Combine all resistive and reactive terms.

3) Convert the total series impedance to polar form.

4) Use V_S and Z_T to calculate the total series current.

5) Use the total series current and component values to calculate the component voltages.

Refer to the series RL circuit and associated phasors in Figure 31-1.

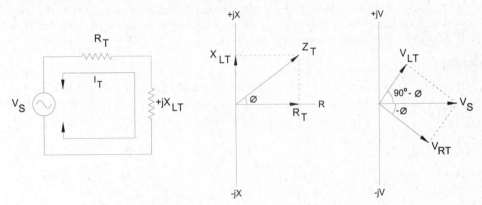

Figure 31-1: Series RL Circuit and Associated Phasors

Step 1: Calculate the inductive reactance for all inductors. The inductive reactance is equal to

$$X_L = 0 + j(2\pi fL) = (2\pi fL) \angle +90°$$

Step 2: Combine all resistive and reactive terms. For a series RL circuit that consists of M resistors and N inductors the total impedance is

$$Z_T = (R_1 + R_2 + ... + R_M) - j(X_{L1} + X_{L2} + ... + X_{LN})$$
$$= R_T + jX_{LT}$$

Step 3: Convert the total series impedance to polar form. For the rectangular form $Z_T = R_T + jX_{LT} = Z_T \angle \varphi$, the amplitude and phase angle are

Amplitude: $Z_T = \sqrt{R_T^2 + X_{LT}^2}$

Phase: $\varphi = \tan^{-1}(X_{LT}/R_T)$

For a series RL circuit the phase angle for the impedance is always a positive value between 0° and 90°. To avoid confusion this discussion will use φ to express a positive angle and $-\varphi$ to express a negative angle.

Step 4: Use V_S and Z_T to calculate the total series current. The voltage does not specify a phase, so assume the voltage phase is 0°. Then

$$I_T = (V_S \angle 0)/(Z_T \angle \varphi)$$

$$= (V_S/Z_T) \angle -\varphi$$

Note that the phase of the current phase is the opposite of the phase of the impedance and is a negative value. The current therefore lags the voltage by φ.

Step 5: Use the total series current and component values to calculate the component voltages. For each resistor

$$V_R = (R \angle 0)(I_T \angle -\varphi)$$

$$= (R \times I_T) \angle -\varphi$$

For each inductor

$$V_L = (X_L \angle 90°)(I_T \angle -\varphi)$$

$$= (X_L \times I_T) \angle (90° -\varphi)$$

Note that the resistor voltage lags the source voltage by $-\varphi$ and that the inductor voltages are 90° ahead of the resistor voltages (i.e., the inductor voltages lead the resistor voltages by 90°). Alternatively the resistor voltages are 90° behind the inductor voltages (i.e., the resistor voltages lag the inductor voltages by 90°). Also, per Kirchhoff's Voltage Law, note that $V_S = V_{RT} + V_{LT}$, as shown by the voltage phasors in Figure 31-1.

Example:

Determine the total current and voltages for the circuit of Figure 31-2 for f = 100 Hz.

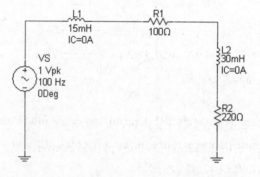

Figure 31-2: Example Series RL Circuit

Step 1: Calculate the inductive reactance for all inductors. For $L1$ = 15 mH and $L2$ = 30 mH

$$X_{L1} = j[(2\pi)(100 \text{ Hz})(15 \text{ mH})] = j(9.42 \ \Omega)$$

$$X_{L2} = j[(2\pi)(100 \text{ Hz})(30 \text{ mH})] = j(18.8 \ \Omega)$$

Step 2: Combine all resistive and reactive terms.

$$Z_T = j(9.4\ \Omega) + 100\ \Omega + j(18.8\ \Omega) + 220\ \Omega$$
$$= (100\ \Omega + 220\ \Omega) + j(9.4\ \Omega + 18.8\ \Omega)$$
$$= 320\ \Omega + j(28.3\ \Omega)$$

Step 3: Convert the total series impedance to polar form. For $Z_T = 320\ \Omega + j(28.3\ \Omega)$

Amplitude: $Z_T = \sqrt{(320\ \Omega)^2 + (28.3\ \Omega)^2} = 321\ \Omega$

Phase: $\varphi = \tan^{-1}(28.3\ \Omega\ /\ 320\ \Omega) = \tan^{-1}(0.0883) = 5.05°$

Polar form: $Z_T = 321\ \Omega \angle 5.05°$

Step 4: Use V_S and Z_T to calculate the total current. $V_S = 1$ Vpk $\angle 0°$ so

$$I_T = V_S\ /\ Z_T$$
$$= (1\ \text{Vpk} \angle 0°)\ /\ (321\ \Omega \angle 5.05°)$$
$$= 3.11\ \text{mApk} \angle -5.05°$$

Step 5: Use the total series current and component values to calculate the component voltages. For each of the components the voltages in polar and rectangular form are

$$V_{L1} = I_T \times X_{C1} = (3.11\ \text{mApk} \angle -5.05°)(9.42\ \Omega \angle 90°)$$
$$= 29.3\ \text{mVpk} \angle 85.0° = 2.58\ \text{mVpk} + j(29.2\ \text{mVpk})$$

$$V_{R1} = I_T \times R1 = (3.11\ \text{mApk} \angle -5.05°)(100\ \Omega \angle 0°)$$
$$= 311\ \text{mVpk} \angle -5.05° = 310\ \text{mVpk} - j(27.4\ \text{mVpk})$$

$$V_{L2} = I_T \times X_{C2} = (3.11\ \text{mApk} \angle -5.05°)(18.8\ \Omega \angle 90°)$$
$$58.7\ \text{mVpk} \angle 85.0° = 5.16\ \text{mVpk} + j(58.4\ \text{mVpk})$$

$$V_{R1} = I_T \times R2 = (3.11\ \text{mApk} \angle -5.05°)(220\ \Omega \angle 0°)$$
$$= 685\ \text{mVpk} \angle -5.05° = 682\ \text{mVpk} - j(60.3\ \text{mVpk})$$

As a check,

$$V_{RT} = V_{R1} + V_{R2}$$
$$= 310\ \text{mVpk} - j(27.4\ \text{mVpk}) + 682\ \text{mVpk} - j(60.3\ \text{mVpk})$$
$$= (310\ \text{mVpk} + 682\ \text{mVpk}) - j(27.4\ \text{mVpk} + 60.3\ \text{mVpk})$$
$$= 992\ \text{mVpk} - j(87.7\ \text{mVpk})$$

$$V_{LT} = V_{L1} + V_{L2}$$
$$= 2.58\ \text{mVpk} + j(29.2\ \text{mVpk}) + 5.16\ \text{mVpk} + j(58.4\ \text{mVpk})$$
$$= (2.58\ \text{mVpk} + 5.16\ \text{mVpk}) + j(29.2\ \text{mVpk} + 58.4\ \text{mVpk})$$
$$= 7.74\ \text{mVpk} + j(87.6\ \text{mVpk})$$

So

$$V_S = V_{RT} + V_{LT}$$
$$= 992\ \text{mVpk} - j(87.7\ \text{mVpk}) + 7.74\ \text{mVpk} + j(87.6\ \text{mVpk})$$
$$= (992\ \text{mVpk} + 7.74\ \text{mVpk}) + j(87.6\ \text{mVpk} - 87.7\ \text{mVpk})$$
$$= 1.000\ \text{Vpk} - j(0.1\ \text{mVpk}) \approx 1.0\ \text{Vpk} + j(0.0\ \text{Vpk}) = 1.0\ \text{Vpk} \angle 0°$$

The voltages comply with Kirchhoff's Voltage Law. The minor deviation is due to rounding in the calculations.

Now, calculate the voltages and currents for the circuit of Figure 31-2 for $f = 1$ kHz and $f = 10$ kHz in both polar and rectangular form and record the values in Table 31-1.

Table 31-1: Calculated Values for Example Series RL Circuit

Circuit Value	$f = 1$ kHz		$f = 10$ kHz	
	Rectangular	Polar	Rectangular	Polar
$X_{L1} = 2\pi(f)(L1)$				
$X_{L2} = 2\pi(f)(L2)$				
$R_T = R1 + R2$				
$X_{LT} = X_{L1} + X_{L2}$				
$Z_T = R_T + X_{LT}$				
$I_T = V_S / Z_T$				
$V_{L1} = I_T \times X_{L1}$				
$V_{L2} = I_T \times X_{L2}$				
$V_{R1} = I_T \times R1$				
$V_{R2} = I_T \times R2$				

31.2.2 Parallel RC Circuits

The basic procedure for analyzing parallel RL circuits is

1) Calculate the inductive reactance for all inductors.

2) Use V_S and the branch resistances and reactances to calculate the branch currents.

3) Add the branch currents to determine the total current.

4) Use V_S and I_T to calculate the total impedance.

Refer to the parallel RL circuit and associated current phasors in Figure 31-3.

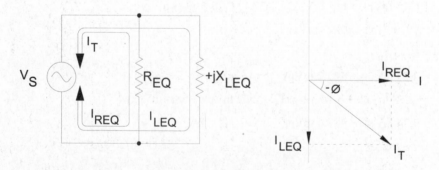

Figure 31-3: Parallel RL Circuit and Associated Phasors

Step 1: Calculate the capacitive reactance for all inductors. The inductive reactance is equal to

$$X_L = 0 + j(2\pi fL) = (2\pi fL) \angle 90°$$

Step 2: Use V_S and the branch resistances and reactances to calculate the branch currents. The voltage does not specify a phase so assume that the voltage phase is 0°. The current for each resistive branch is

$$I_R = (V_S \angle 0°) / (R \angle 0°)$$

$$= (V_S / R) \angle 0° = (V_S / R) + \mathrm{j}(0 \text{ A})$$

The current for each reactive branch is

$$I_L = (V_S \angle 0°) / (X_L \angle 90°)$$

$$= (V_S / X_L) \angle -90° = 0 \text{ A} - \mathrm{j}(V_S / X_L)$$

Step 3: Add the branch currents to determine the total current. From Step 2 this is

$$I_T = (V_S / R_1) + \ ... + (V_S / R_M) - \mathrm{j}[(V_S / X_{L1}) + ... + (V_S / X_{LN})]$$

$$= V_S(1/R_1 + ... + 1/R_M) - \mathrm{j}[V_S(1/X_{L1} + ... + 1/X_{LN})]$$

If we let $R_{EQ} = 1 / (1/R_1 + ... + 1/R_M)$ and $X_{EQ} = 1/ (1/X_{L1} + ... + 1/X_{LN})$ this becomes

$$I_T = (V_S / R_{EQ}) - \mathrm{j}(V_S / X_{EQ})$$

$$= I_{REQ} - \mathrm{j}(I_{XEQ})$$

Amplitude: $I_T = \sqrt{(I_{REQ})^2 + (I_{XEQ})^2}$

Phase: $-\varphi = \tan^{-1}(-I_{XEQ} / I_{REQ})$

For a parallel RL circuit the phase angle for the current is always a negative value between 0° and −90°.

Step 4: Use V_S and I_T to calculate the total impedance. Then

$$Z_T = (V_S \angle 0) / (I_T \angle -\varphi)$$

$$= V_S / I_T \angle (0 + \varphi)$$

$$= (V_S / I_T) \angle \varphi$$

Note that the impedance phase is the opposite of the current phase and is positive.

Example:

Determine the currents and total impedance for the circuit of Figure 31-4 for f = 100 Hz.

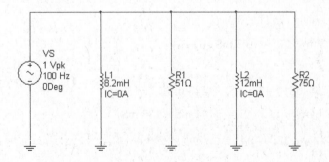

Figure 31-4: Example Parallel RL Circuit

Step 1: Calculate the inductive reactance for all inductors. For $L1$ = 8.2 mH and $L2$ = 12 mH

$$X_{L1} = \mathrm{j}[(2\pi)(100 \text{ Hz})(8.2 \text{ mH})] = \mathrm{j}(5.15 \ \Omega) = 5.15 \ \Omega \angle 90°$$

$$X_{L2} = \mathrm{j}[(2\pi)(100 \text{ Hz})(12 \text{ mH})] = \mathrm{j}(7.54 \ \Omega) = 7.54 \ \Omega \angle 90°$$

Step 2: Use V_S and the branch resistances and reactances to calculate the branch currents.

$$I_{L1} = (1.0 \text{ Vpk} \angle 0°) / (5.15 \text{ k}\Omega \angle 90°)$$
$$= 194 \text{ mApk} \angle -90° = -j(194 \text{ mApk})$$

$$I_{R1} = (1.0 \text{ Vpk} \angle 0°) / (51 \Omega \angle 0°)$$
$$= 19.6 \text{ mA} \angle 0° = 19.6 \text{ mApk}$$

$$I_{L2} = (1.0 \text{ Vpk} \angle 0°) / (7.54 \Omega \angle 90°)$$
$$= 133 \text{ mApk} \angle -90° = -j(133 \text{ mApk})$$

$$I_{R2} = (1.0 \text{ Vpk} \angle 0°) / (75 \Omega \angle 0°)$$
$$= 13.3 \text{ mA} \angle 0° = 13.3 \text{ mApk}$$

Step 3: Add the branch currents to determine the total current. The rectangular and polar forms are

$$I_T = -j(194 \text{ mApk}) + 19.6 \text{ mApk} - j(133 \text{ mApk}) + 13.3 \text{ mApk}$$
$$= (19.6 \text{ mApk} + 13.3 \text{ mApk}) - j(194 \text{ mApk} + 133 \text{ mApk})$$
$$= 32.9 \text{ mApk} - j(327 \text{ mApk}) = 328 \text{ mA} \angle -84.2°$$

Step 4: Use V_S and I_T to calculate the total impedance. $V_S = 1 \text{ Vpk} \angle 0°$ so

$$Z_T = V_S / I_T$$
$$= (1 \text{ Vpk} \angle 0°) / (328 \text{ mApk} \angle -84.2°)$$
$$= 3.05 \Omega \angle 84.2°$$

As a check, determine Z_T by calculating the reciprocal of the sum of the reciprocals for each resistance and reactance.

$$1 / X_{L1} = (1 \angle 0°) / (5.15 \Omega \angle 90°)$$
$$= 194 \text{ mS} \angle -90° = -j(194 \text{ mS})$$

$$1 / R1 = (1 \angle 0°) / (51 \Omega \angle 0°)$$
$$= 19.6 \text{ mS} \angle 0° = 19.6 \text{ mS}$$

$$1 / X_{L2} = (1 \angle 0°) / (7.54 \Omega \angle 90°)$$
$$= 133 \text{ mS} \angle -90° = -j(133 \text{ mS})$$

$$1 / R2 = (1 \angle 0°) / (75 \Omega \angle 0°)$$
$$= 13.3 \text{ mS} \angle 0° = 13.3 \text{ mS}$$

From this

$$1 / Z_T = -j(194 \text{ mS}) + 19.6 \text{ mS} - j(133 \text{ mS}) + 13.3 \text{ mS}$$
$$= (19.6 \text{ mS} + 13.3 \text{ mS}) - j(194 \text{ mS} + 133 \text{ mS})$$
$$= 32.9 \text{ mS} - j(326 \text{ mS}) = 328 \text{ mS} \angle -84.2°$$

Therefore

$$Z_T = (1 \angle 0°) / (328 \text{ mS} \angle -84.2°)$$
$$= 3.05 \Omega \angle 84.2°$$

The total impedance values check.

Now, calculate the currents and total impedance for the circuit of Figure 31-4 for f = 1 kHz and f = 10 kHz in both polar and rectangular form and record the values in Table 31-2.

Table 31-2: Calculated Values for Example Parallel RL Circuit

Circuit Value	f = 1 kHz		f = 10 kHz	
	Rectangular	Polar	Rectangular	Polar
$X_{L1} = 2\pi(f)(L1)$				
$X_{L2} = 2\pi(f)(L2)$				
$I_{R1} = V_S / R1$				
$I_{R2} = V_S / R2$				
$I_{L1} = V_S / X_{L1}$				
$I_{L2} = V_S / X_{L2}$				
$I_T = I_{R1} + I_{R2} + I_{L1} + I_{L2}$				
$Z_T = V_S / I_T$				

31.3 Design Verification

31.3.1   Series RL Circuit

1. Open the circuit Ex31-01.

2. Select the two-channel oscilloscope from the Instrument toolbar.

3. Connect the "+" input of Channel A between *VS* and *L1* and the "–" input to circuit ground.

4. Connect the inputs of Channel B across *L1* with the "+" input closer to the positive terminal of *VS*.

5. Change the color of the "+" input of Channel B to a different color from the rest of the circuit wiring. This is not necessary but will help you to distinguish between the waveforms.

6. Click the **Run** switch to simulate the circuit, run the simulation to run for at least 50 ms, and click the **Run** switch to stop the simulation.

7. Adjust the "Scale" values of both channels so that each waveform is between four and six divisions high.

8. Adjust the timebase scale division so that no more than one cycle of each waveform is visible.

9. Adjust the scroll bar below the oscilloscope display so that you can see one positive peak for both waveforms. Only one positive peak of each should be visible.

10. Right-click Cursor 1 and select "Select Trace ID" from the cursor right-click menu.

11. Select "Channel A" from the drop-down list (if it is not already selected) and click the **OK** button.

12. Right-click Cursor 1 and select "Go to next Y_MAX =>" from the cursor right-click menu. The cursor should move to the positive peak of Channel A.

13. Right-click Cursor 2 and select "Select Trace ID" from the cursor right-click menu.

14. Select "Channel B" from the drop-down list (if it is not already selected) and click the **OK** button.

15. Right-click Cursor 2 and select "Go to next Y_MAX <=" from the cursor right-click menu. The cursor should move to the positive peak of Channel B.

16. Record the Channel B amplitude in Table 31-3 as the value of V_{L1} for f = 100 Hz.

17. Use the value of $T2 - T1$ to calculate the phase. Record this phase in Table 31-3 as the value of φ_{L1} for the "f = 100 Hz" column.

18. Repeat Steps 4 through 17 for $R1$, $L2$, and $R2$ and record the amplitude and phase values in Table 31-3.

19. Change the frequency of the ac voltage source to 1 kHz and 10 kHz and repeat Steps 6 through 18.

Table 31-3: Measured Values for Example Series RL Circuit

Circuit Value	f = 100 Hz	f = 1 kHz	f = 10 kHz
V_{L1}			
φ_{L1}			
V_{R1}			
φ_{R1}			
V_{L2}			
φ_{L2}			
V_{R2}			
φ_{R2}			

Do your measured values in Table 31-3 agree with your calculated values in Table 31-1?

31.3.2 Parallel RL Circuit

You cannot directly measure currents for the example parallel RL circuit in Figure 31-4. Because the currents and voltages for a resistor are in phase, however, you can measure the voltage across a small sense resistor that is in series with each parallel branch and use the measured voltage to calculate the current. You must, however, ensure that the sense resistance is no more than 1/100th the impedance of the branch. For this circuit R = 1 mΩ is acceptable. Note that $R1$ and $R2$ are the sense resistors for their branches.

1. Open the circuit Ex31-02.

2. Select the two-channel oscilloscope from the Instrument toolbar.

3. Connect the "+" input of Channel A between VS and $L1$ and the "–" input to circuit ground.

4. Connect the "+" input of Channel B between $L1$ and $RS1$ and the "–" input to circuit ground.

5. Change the color of the "+" input of Channel B to a different color from the rest of the circuit wiring. This is not necessary but will help you to distinguish between the waveforms.

6. Click the **Run** switch to simulate the circuit, run the simulation to run for at least 50 ms, and click the **Run** switch to stop the simulation.

7. Adjust the "Scale" values of both channels so that each waveform is between four and six divisions high.

8. Adjust the timebase scale division so that no more than one cycle of each waveform is visible.

9. Adjust the scroll bar below the oscilloscope display so that you can see one positive peak for both waveforms. Only one positive peak of each should be visible.

10. Right-click Cursor 1 and select "Select Trace ID" from the cursor right-click menu.

11. Select "Channel A" from the drop-down list (if it is not already selected) and click the **OK** button.

12. Right-click Cursor 1 and select "Go to next Y_MAX =>" from the cursor right-click menu. The cursor should move to the positive peak of Channel A.

13. Right-click Cursor 2 and select "Select Trace ID" from the cursor right-click menu.

14. Select "Channel B" from the drop-down list (if it is not already selected) and click the **OK** button.

15. Right-click Cursor 2 and select "Go to next Y_MAX <=" from the cursor right-click menu. The cursor should move to the positive peak of Channel B.

16. Record the Channel B amplitude divided by the sense resistance in Table 31-4 as the value of I_{L1} for f = 100 Hz.

17. Use the value of $T2 - T1$ to calculate the phase. Record this phase in Table 31-4 as the value of φ_{L1} for the "f = 100 Hz" column.

18. Repeat Steps 4 through 17 for $R1$, $RS2$ and $L2$, and $R2$ and record the amplitude and phase values in Table 31-4.

19. Change the frequency of the ac voltage source to 1 kHz and 10 kHz and repeat Steps 6 through 18.

Table 31-4: Measured Values for Example Parallel RL Circuit

Circuit Value	f = 100 Hz	f = 1 kHz	f = 10 kHz
I_{L1}			
φ_{L1}			
I_{R1}			
φ_{R1}			
I_{L2}			
φ_{L2}			
I_{R2}			
φ_{R2}			

Do your measured values in Table 31-4 agree with your calculated values in Table 31-2?

31.4 Application Exercise

The ac analysis of series-parallel RL circuits is similar to the ac analysis of series-parallel RC circuits, and combines the ac analysis techniques for series RL circuits and parallel RL circuits. You replace RL combinations in the circuit with complex circuit equivalents until you determine the total circuit impedance, and then work backwards to find the voltages and currents for individual circuit components.

Refer to the series-parallel RL circuit in Figure 31-5.

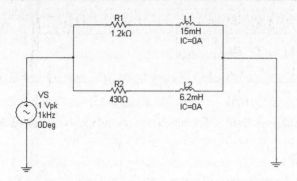

Figure 31-5: Example Series-Parallel RL Circuit

Calculate each of the values shown in Table 31-5 for the example series-parallel RL circuit.

Table 31-5: Calculated Values for Series-Parallel RL Circuit

Circuit Value	Rectangular Form	Polar Form
$X_{L1} = 2\pi(1\ kHz)(100\ nF)$		
$X_{L2} = 2\pi(1\ kHz)(470\ nF)$		
$Z_{EQ1} = R1 + X_{L1}$		
$Z_{EQ2} = R2 + X_{L2}$		
$I_{R1} = I_{L1} = VS / Z_{EQ1}$		
$I_{R2} = I_{L2} = VS / Z_{EQ2}$		
$I_T = I_{R1} + I_{R2}$		
$Z_T = VS / I_T$		
$V_{R1} = I_{R1} \times R1$		
$V_{L1} = I_{L1} \times X_{L1}$		
$V_{R2} = I_{R2} \times R2$		
$V_{L2} = I_{L2} \times X_{L2}$		

Open the circuit file Pr31-01. Determine the amplitude and phase for each of the values shown in Table 31-6.

Table 31-6: Measured Values for Example Series-Parallel RL Circuit

Circuit Value	Measured (Polar Form)
I_T	
V_{L1}	
I_{L1}	
V_{R1}	
I_{R1}	
V_{L2}	

Circuit Value	Measured (Polar Form)
I_{L2}	
V_{R2}	
I_{R2}	

Do your measured values in Table 31-6 match your calculated values in Table 31-5?

31.5 Section Summary

This circuit covered the ac analysis of series, parallel, and series-parallel RL circuits. Ac analysis requires an oscilloscope to determine both the amplitude and phase components of each circuit element. This limits direct circuit measurement to ac voltages, but circuit and sense resistors allow you to determine the amplitude and phase of circuit currents.

32. AC Response of RLC Circuits

32.1 Introduction

The ac analysis of RLC circuits is identical to that for RC and RL circuits. The only difference is that the circuit will contain both capacitive and inductive reactance. The effect of the capacitive reactance is for current to lead the voltage by 90° while the effect of the inductive reactance is for voltage to lead the current by 90°. A common mnemonic for this is "ELI the ICE man." ELI is a reminder that electromotive force or voltage (E) comes before, or leads, current (I) for inductors (L). ICE is a reminder that current (I) comes before, or leads, electromotive force or voltage (E) for capacitors (C).

Refer to the phasor diagrams in Figure 32-1.

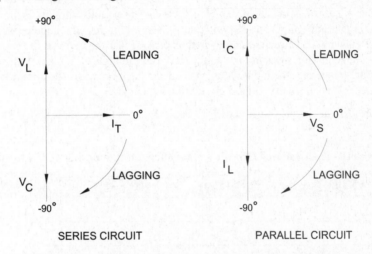

Figure 32-1: Series and Parallel Circuit Reactive Phasors

For a series circuit $I_C = I_L = I_T$. Since V_L leads I_T by 90° and V_C lags I_T by 90°, V_L and V_C are 180° out of phase and therefore oppose each other. Consequently, the net reactive voltage is the difference between V_L and V_C. For a parallel circuit $V_C = V_L = V_S$. Since I_L lags V_S by 90° and I_C leads V_S by 90°, I_L and I_C are 180° out of phase and therefore oppose each other. Consequently the net reactive current is the difference between I_L and I_C.

In this section you will:

- Investigate some electrical characteristics of series, parallel, and series-parallel RLC circuits.
- Use the Multisim software to verify the results of your circuit analysis of RLC circuits.

32.2 Pre-Lab

32.2.1 Series RLC Circuit

Refer to the series RLC circuit of Figure 32-2.

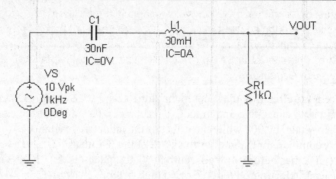

Figure 32-2: Example Series RLC Circuit

At low frequencies *L1* is a short so that the circuit acts like a series RC circuit. From the phasor diagram for a series RC circuit in Figure 30-1, V_{R1} will lead *VS*. A circuit for which the output voltage leads the input voltage is called a lead circuit. At high frequencies *C1* is a short so that the circuit act like a series RL circuit. From the phasor diagram for a series RL circuit in Figure 31-1 V_{R1} will lag *VS*. A circuit for which the output voltage lags the input voltage is called a lag circuit. As the circuit frequency increases the circuit in Figure 32-2 changes from a lead circuit to a lag circuit.

Calculate and record the circuit values shown in Table 32-1 for the example series RLC circuit in Figure 32-2.

Table 32-1: Calculated Circuit Values for Example Series RLC Circuit

Circuit Value	f = 500 Hz	f = 5 kHz	f = 50 kHz
X_{L1}			
X_{C1}			
$Z_T = R1 + \text{j}(X_{L1} - X_{C1})$			
$I_T = VS / Z_T$			
$VOUT = I_T \times R1$			

Does the circuit change from a lead circuit to a lag circuit as the frequency increases from 500 Hz to 50 kHz?

32.2.2   Parallel RLC Circuit

Refer to the parallel RLC circuit of Figure 32-3.

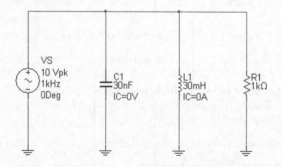

Figure 32-3: Example Parallel RCL Circuit

At low frequencies $L1$ is a low impedance to ground so that the value of I_T is high. At high frequencies $C1$ is a low impedance to ground so that the value of I_T is also high. At some frequency in between, the branch currents of $C1$ and $L1$ cancel so that $I_T = I_{R1}$. As the circuit frequency increases, the circuit impedance changes from a low value to $R1$ and then decreases again so that the total current decreases to some minimum and increases again.

Calculate and record the circuit values shown in Table 32-2 for the example parallel RLC circuit in Figure 32-3.

Table 32-2: Calculated Circuit Values for Example Parallel RLC Circuit

Circuit Value	f = 500 Hz	f = 5 kHz	f = 50 kHz
X_{L1}			
X_{C1}			
$I_{R1} = VS / R1$			
$I_{L1} = VS / X_{L1}$			
$I_{C1} = VS / X_{C1}$			
$I_T = I_{R1} + j(I_{C1} - I_{L1})$			
$Z_T = VS / I_T$			

Does the total circuit impedance increase and then decrease so that the current decreases and then increases again as the frequency increases from 500 Hz to 50 kHz?

32.2.3 Series-Parallel RLC Circuit

Refer to the series-parallel circuit of Figure 32-4.

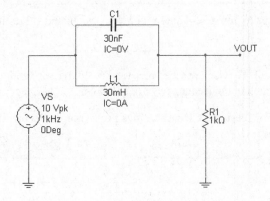

Figure 32-4: Example Series-Parallel RLC Circuit

At low frequencies $L1$ is a low impedance so that most of the source voltage appears across $R1$ as $VOUT$. At high frequencies $C1$ is a low impedance so that most of the source voltage again appears across $R1$ as $VOUT$. At some frequency in between, the impedance of the parallel combination of $C1$ and $L1$ drops appreciable voltage so that the value of $VOUT$ decreases. As the circuit frequency increases, the output voltage decreases and then increases again.

Calculate and record the circuit values shown in Table 32-3 for the example series parallel RLC circuit in Figure 32-4.

Table 32-3: Calculated Circuit Values for Example Series-Parallel RLC Circuit

Circuit Value	f = 500 Hz	f = 5 kHz	f = 50 kHz
X_{L1}			
X_{C1}			
$Z_{EQ} = X_{L1} \| X_{C1}$			
$Z_T = R1 + Z_{EQ}$			
$I_T = VS / Z_T$			
$VOUT = V_{R1} = I_T \times R1$			

Does the output voltage decrease and then increase again as the frequency increases from 500 Hz to 50 kHz?

32.3 Design Verification

32.3.1 Series RLC Circuit

Open the circuit file Ex32-01. Measure and record the amplitude and phase of the output voltage across *R1* for each of the frequencies in Table 32-4.

Table 32-4: Measured Circuit Values for Example Series RLC Circuit

Circuit Value	f = 500 Hz	f = 5 kHz	f = 50 kHz
V_{R1} Amplitude			
V_{R1} Phase			

Do your measured values in Table 32-4 agree with your calculated values in Table 32-1?

32.3.2 Parallel RLC Circuit

Open the circuit file Ex32-02. Use the sense resistor *RS1* to measure and record the amplitude and phase of the total current for each of the frequencies in Table 32-5.

Table 32-5: Measured Circuit Values for Example Series RLC Circuit

Circuit Value	f = 500 Hz	f = 5 kHz	f = 50 kHz
I_T Amplitude			
I_T Phase			

Do your measured values in Table 32-5 agree with your calculated values in Table 32-2?

32.3.3 Series-Parallel RLC Circuit

Open the circuit file Ex32-03. Measure and record the amplitude and phase of the output voltage across *R1* for each of the frequencies in Table 32-6.

Table 32-6: Measured Circuit Values for Example Series RLC Circuit

Circuit Value	$f = 500$ Hz	$f = 5$ kHz	$f = 50$ kHz
V_{R1} Amplitude			
V_{R1} Phase			

Do your measured values in Table 32-6 agree with your calculated values in Table 32-3?

32.4 Application Exercise

For each of the RLC circuits you investigated you should have seen that some circuit characteristic appeared to reach some maximum or minimum value at some frequency. For the series RLC circuit the value of **VOUT** was maximum and phase was minimum when the value of I_T was maximum. This occurred when the values of X_{L1} and X_{C1} were equal and cancelled each other out, so that the value of Z_T was minimum and about equal to the value of **R1**. For the parallel RLC circuit I_T was minimum when I_{L1} and I_{C1} were equal and cancelled each other out. This occurred when $VS / X_{L1} = VS / X_{C1}$ so that the values of X_{L1} and X_{C1} again were equal. For the values of X_{L1} and X_{C1} to be equal

$X_{L1} = X_{C1}$

$2\pi(f)(L1) = 1 / [2\pi(f)(C1)]$

$(2\pi f)^2 = 1 / [(L1)(C1)]$

$f^2 = 1 / [(2\pi)^2(L1)(C1)]$

$f = 1 / [2\pi\sqrt{(L1)(C1)}]$

For the example circuits **L1** = 30 mH and **C1** = 30 nF so

$f = 1 / [2\pi\sqrt{(30\text{ mH})(30\text{ nF})}] = 1 / [2\pi\sqrt{900 \times 10^{-12}}] = 1 / [(2\pi)(30 \times 10^{-6})] = 5.31$ kHz

For the circuit in Figure 32-5, calculate the frequency for which the value of **VOUT** is maximum. Then open the circuit Pr32-01 and use the ac multimeter to measure the value of **VOUT** for each of the frequencies in Table 32-7.

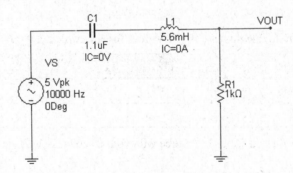

Figure 32-5: Practice Series RLC Circuit

Table 32-7: Measured Output Voltage Values for Practice Series RLC Circuit

f	VOUT (Vrms)	f	VOUT (Vrms)
10 Hz		5 kHz	
20 Hz		10 kHz	
50 Hz		20 kHz	
100 Hz		50 kHz	
200 Hz		100 kHz	
500 Hz		200 kHz	
1 kHz		500 kHz	
2 kHz		1 MHz	

Does the value of *VOUT* maximize at the calculated frequency?

32.5 Section Summary

This section investigated some of the characteristics of series, parallel, and series-parallel RLC circuits. For each RLC circuit there is a unique frequency for which the inductive and capacitive reactances are equal and for which some characteristics of the circuit reach a maximum or minimum value. This frequency, which depends on the capacitance and inductance values, is called the **resonance** or **natural frequency** of the circuit. You will learn more about resonant circuits in future sections.

33. Power in Complex Circuits

33.1 Introduction

The total impedance of a complex circuit consists of resistance and reactance. The resistance in the circuit dissipates energy at a certain rate as **real** or **true power** (P_T), while the reactance stores and returns energy at a certain rate as **reactive power** (P_R). These resistive and reactive components combine to form the total power, called **apparent power** (P_A). For a complex circuit with total impedance $Z_T = R_T \pm jX_T$, if V_{RMS} is the total rms voltage and I_{RMS} is the total rms current then

$$P_A = V_{RMS} \times I_{RMS} = I_{RMS}^2 \times Z_T = V_{RMS}^2 / Z_T$$

$$P_T = I_{RMS}^2 \times R_T = V_{RT}^2 / R_T$$

$$P_R = I_{RMS}^2 \times X_T = V_{XT}^2 / X_T$$

The units of apparent power are volt-amps (VA), the units of real power in watts (W), and the units of reactive power are volt-amps reactive (VAR).

Just as the impedance triangle illustrates the relationship between resistance, reactance, and impedance, the power triangle illustrates the relationship between real power, reactive power, and apparent power. Refer to Figure 33-1.

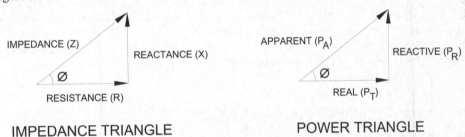

IMPEDANCE (Z)   REACTANCE (X)   APPARENT (P_A)   REACTIVE (P_R)

Ø   Ø

RESISTANCE (R)   REAL (P_T)

IMPEDANCE TRIANGLE   POWER TRIANGLE

Figure 33-1: Impedance and Power Triangles for Complex Circuits

The angle φ for both triangles is the phase angle for the circuit. From basic trigonometry

$$P_A = \sqrt{P_T^2 + P_R^2}$$

$$P_T = P_A \cos \varphi$$

$$P_R = P_A \sin \varphi$$

The term **cos φ** is of special interest and called the **power factor** (PF), because multiplying the apparent power by the cos φ gives the real power in the circuit. When the circuit is purely resistive, $\varphi = 0$, cos $\varphi = 1$, and $P_T = P_A$. The Multisim wattmeter shows the power factor so that you can determine the apparent and reactive power when you measure power in a complex circuit.

To calculate the apparent power from the real power and power factor use the equation

$$P_A = P_T / P_F$$

To calculate the reactive power from the real power and power factor you can use the equation

$$P_R = (P_T / P_F) \sin [\cos^{-1} (PF)]$$

In this section you will:

- Calculate the real, reactive, and apparent powers in complex circuits and investigate the relationship between them.

- Use the Multisim software to verify the properties of power in complex circuits.

33.2 Pre-Lab

Refer to the complex circuit shown in Figure 33-2.

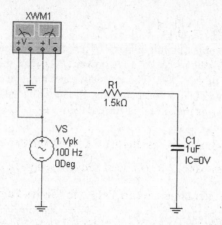

Figure 33-2: Example Complex Circuit Power Measurement

To calculate power directly the voltage and currents must be rms values.

$$VS = 1.0 \text{ Vpk} = 0.707 \times 1.0 = 0.707 \text{ V}_{RMS}$$

$$X_{C1} = 1/ [2\pi(100 \text{ Hz})(1 \text{ }\mu\text{F})] = 1.59 \text{ k}\Omega$$

$$|Z_T| = \sqrt{(1.5 \text{ k}\Omega)^2 + (1.59 \text{ k}\Omega)^2} = 2.19 \text{ k}\Omega$$

$$\varphi = -\tan^{-1}(1.59 \text{ k}\Omega / 1.5 \text{ k}\Omega) = -46.7°$$

$$I_T = VS / Z_T = 0.707 \text{ V}_{RMS} / 2.19 \text{ k}\Omega = 323 \text{ }\mu\text{A}_{RMS}$$

$$P_A = VS \times I_T = (0.707 \text{ V}_{RMS})(323 \text{ }\mu\text{A}_{RMS}) = 229 \text{ }\mu\text{VA}$$

$$P_T = I_T^2 \times R1 = (323 \text{ }\mu\text{A}_{RMS})^2 (1.5 \text{ k}\Omega) = 157 \text{ }\mu\text{W}$$

$$P_R = I_T^2 \times X_{C1} = (323 \text{ }\mu\text{A}_{RMS})^2 (1.59 \text{ k}\Omega) = 166 \text{ }\mu\text{VAR}$$

As a check, the power factor is $\cos \varphi = \cos (-46.7°) = 0.686$, so

$$P_R = (229 \text{ }\mu\text{VA})(0.686) = 157 \text{ }\mu\text{W}$$

which agrees with the calculation of $I_T^2 \times R1$. As a second check the apparent power should be equal to $I_T^2 \times Z_T$ so

$$P_A = (323 \text{ }\mu\text{A}_{RMS})^2 (2.19 \text{ k}\Omega) = 229 \text{ }\mu\text{VA}$$

which agrees with the calculation of $VS \times I_T$.

Open the circuit file Ex33-01. Simulate the circuit and verify that the measured real power and power factor values agree with the calculations above.

Refer to the circuit of Figure 33-3.

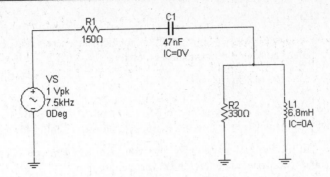

Figure 33-3: Series-Parallel RLC Circuit for Power Calculations

Calculate each of the series-parallel RLC circuit values listed in Table 33-1.

Table 33-1: Calculated Circuit Values for Series-Parallel RLC Circuit

Circuit Value	Rectangular Form	Polar Form
VS (rms) = 0.707 × VS		
X_{C1} = 1 / [2π(f)($C1$)]		
X_{L1} = 2π(f)($L1$)		
Z_{EQ1} = $R1$ + X_{C1}		
Z_{EQ2} = $R2$ ‖ $L1$		
Z_T = Z_{EQ1} + Z_{EQ2}		
R_T		
X_T		
φ = $\tan^{-1}$ (X_T / R_T)		
PF = cos φ		
I_T = I_{R1} = I_{C1} = I_{ZEQ2} = VS (rms) / Z_T		
P_{ZT} = P_A = I_T × VS		
P_{ZEQ1} = I_T^2 × Z_{EQ1}		
P_{ZEQ2} = I_T^2 × Z_{EQ2}		

Open the circuit file Ex33-02. Use the wattmeter to measure the power factor PF and true power P_T for the circuit values listed in Table 33-2. Then calculate and record the apparent and reactive powers for each.

Table 33-2: Measured Circuit Values for Series-Parallel RLC Circuit

Circuit Value	PF	P_T	$P_A = P_T / PF$	$P_R = P_A$ sin [$\cos^{-1}$ (PF)]
Z_T				
Z_{EQ1}				
Z_{EQ2}				

Do your calculated values in Table 33-1 agree with your measured and calculated values in Table 33-2?

33.3 Application Exercise

As the application exercise in Section 32 showed, the reactances in a series RLC circuit will cancel at a special frequency f_r, called the resonant frequency, given by

$$f_r = 1/(2\pi\sqrt{LC})$$

When the reactances cancel the circuit is purely resistive so that $PF = 1$. You can use the wattmeter to determine the resonant frequency by varying the frequency and determining for which $PF = 1$.

Open the practice circuit file Pr33-01 and connect the wattmeter to the circuit. For each frequency f in Table 33-3 measure and record the power factor PF for the circuit.

Table 33-3: Measured Power Factor for Practice Circuit

f	PF	f	PF
500 Hz		3.0 kHz	
1.0 kHz		3.5 kHz	
1.5 kHz		4.0 kHz	
2.0 kHz		4.5 kHz	
2.5 kHz		5.0 kHz	

What is the approximate resonant frequency f_r?

What is the approximate value of $L1 = 1 / [(2\pi f_r)^2(C1)]$?

33.4 Section Summary

This section examined power in complex circuits and investigated the relationship between apparent, real, and reactive power. You can find the apparent power (P_A) by multiplying the rms voltage by the rms current and determine the real power (P_T) by multiplying the apparent power by the power factor (PF), which is equal to the cosine of the phase angle ($\cos\varphi$). Just as you can find dc power in resistors by using the equations $P = I^2R = V^2/R$, you can find the power in ac circuits by using the formulas $P_T = I^2R = V^2/R$, $P_R = I^2X = V^2/X$, and $P_A = I^2Z = V^2/Z$ provided that the voltage and current are rms values.

34. Series Resonance

34.1 Introduction

As you now know, the total impedance of a series RLC circuit is

$$Z_T = R_T + j(X_{LT} - X_{LC})$$

where

R_T is the total series resistance,

X_{LT} is the total series inductive reactance, and

X_{CT} is the total series capacitive reactance.

When $X_{LT} = X_{CT}$ the imaginary term $j(X_{LT} - X_{CT}) = 0$ so that $Z_T = R_T$. This circuit condition for a series RLC circuit is called series resonance. At resonance the total impedance is minimum, the total series current is maximum, and the phase angle $\varphi = 0°$.

In this section you will:

- Verify the resonant frequency for series RLC circuits.

- Investigate the behavior of series RLC circuits at resonance.

- Learn to use the Multisim Bode plotter.

- Use the Multisim software to verify your analysis of series RLC circuits at resonance.

34.2 Pre-Lab

Refer to the series RLC circuit in Figure 34-1.

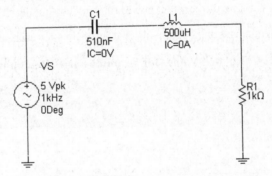

Figure 34-1: Example Series Resonant Circuit

The resonant frequency f_r for this circuit is

$$f_r = \frac{1}{2\pi\sqrt{(L1)(C1)}} = \frac{1}{2\pi\sqrt{(510\text{ nF})(500\text{ μH})}} = 9.97\text{ kHz}$$

At this frequency

$$X_{C1} = -j\,(1\,/\,[2\pi(f_r)(L1)])$$

$$= -j(1\,/\,[(2\pi)(9.97\text{ kHz})(510\text{ nF})])$$

$$= -j(31.1\ \Omega) = 31.1\ \Omega\ \angle\ -90°$$

$$X_{L1} = j[2\pi(f_r)(L1)]$$

233

$$= j[(2\pi)(9.97 \text{ kHz})(500 \text{ }\mu\text{H})]$$

$$= j(31.1 \text{ }\Omega) = 31.1 \text{ }\Omega \angle 90°$$

So

$$Z_T = R1 + j(X_{L1} - X_{C1})$$

$$= 1 \text{ k}\Omega + j(31.1 \text{ }\Omega - 31.1 \text{ }\Omega)$$

$$= 1 \text{ k}\Omega + j(0 \text{ }\Omega) = 1 \text{ k}\Omega \angle 0°$$

and

$$I_T = VS / Z_T$$

$$= (5 \text{ Vpk} \angle 0°) / (1 \text{ k}\Omega \angle 0°)$$

$$= 5.0 \text{ mApk} \angle 0° = 3.54 \text{ mA}_{RMS} \angle 0°$$

The voltages across each of the components is then

$$I_{C1} = I_T \times X_{C1}$$

$$= (5.0 \text{ mApk} \angle 0°)(31.1 \text{ }\Omega \angle -90°)$$

$$= 156 \text{ mVpk} \angle -90° = 110 \text{ mV}_{RMS} \angle -90°$$

$$I_{L1} = I_T \times X_{L1}$$

$$= (5.0 \text{ mApk} \angle 0°)(31.1 \text{ }\Omega \angle 90°)$$

$$= 156 \text{ mVpk} \angle 90° = 110 \text{ mV}_{RMS} \angle 90°$$

$$I_{R1} = I_T \times R1$$

$$= (5.0 \text{ mApk} \angle 0°)(1 \text{ k}\Omega \angle 0°)$$

$$= 5.0 \text{ Vpk} \angle 0° = 3.54 \text{ V}_{RMS} \angle 0°$$

Calculate the resonant frequencies and circuit values at resonance for each set of values in Table 34-1.

Table 34-1: Calculated Values for Example Series Resonant Circuit

Circuit Value	$C1 = 220$ nF $L1 = 4.7$ mH $R1 = 1$ kΩ	$C1 = 220$ nF $L1 = 4.7$ mH $R1 = 2$ kΩ	$C1 = 360$ nF $L1 = 420$ μH $R1 = 510$ Ω	$C1 = 910$ nF $L1 = 27$ mH $R1 = 1.5$ kΩ
f_r	9.97 kHz			
X_{C1}	31.1 Ω ∠ −90°			
X_{L1}	31.1 Ω ∠ 90°			
$Z_T = R1 + j(X_{L1} - X_{C1})$	1 kΩ ∠ 0°			
$I_T = (3.54 \text{ V}_{RMS} \angle 0°) / Z_T$	3.54 mA$_{RMS}$ ∠ 0°			
$V_{C1} = I_T \times X_{C1}$	110 mV$_{RMS}$ ∠ −90°			
$V_{L1} = I_T \times X_{L1}$	110 mV$_{RMS}$ ∠ 90°			
$V_{R1} = I_T \times R1$	3.54 V$_{RMS}$ ∠ 0°			

34.3 Design Verification

34.3.1 Resonant Circuit Values

Open the circuit file Ex34-01. Set the values of *C1*, *L1*, and *R1* to each set of values shown in Table 34-2 and set the frequency of the ac voltage source to the corresponding resonant frequency you calculated in the Pre-Lab section. Use a multimeter to measure V_{C1}, V_{L1}, and V_{R1} and record the values in the appropriate column of Table 34-2.

Table 34-2: Measured Values for Example Series Resonant Circuit

Circuit Value	C1 = 510 nF L1 = 500 μH R1 = 1 kΩ	C1 = 220 nF L1 = 4.7 mH R1 = 2 kΩ	C1 = 360 nF L1 = 420 μH R1 = 510 Ω	C1 = 910 nF L1 = 27 mH R1 = 1.5 kΩ
f_r	9.97 kHz			
V_{C1}				
V_{L1}				
V_{R1}				

Do your measured voltage values in Table 34-2 agree with your calculated values in Table 34-1?

34.3.2 The Multisim Bode Plotter

The output of a circuit as a function of frequency is the circuit's **frequency response**. One way to determine the frequency response of a circuit is to manually change the frequency of the voltage source and make circuit measurements for each frequency setting. The Multisim Bode plotter offers a much simpler way to determine a circuit's frequency response. Refer to Figure 34-2 and Figure 34-3.

Figure 34-2: Bode Plotter Tool

Figure 34-3: Bode Plotter Minimized and Enlarged Views

The Bode plotter, which resembles an oscilloscope, is not a real instrument. Its name comes from "Bode plot," which is a graphical representation of the magnitude and phase of a frequency-dependent circuit as function of frequency.

The Bode plotter settings consists of five sections.

1) **Bode plotter graphical display**

The Bode plotter graphical display is the large area on the left side of the instrument. The display, shown in Figure 34-4, is similar to an oscilloscope display.

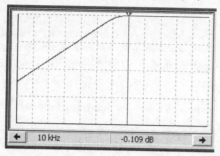

Figure 34-4: Bode Plotter Graphical Display

The Bode plotter magnitude and phase displays each possess one cursor so that you can select specific points on the Bode plot. The status bar below the display provides the magnitude (or phase) of the selected point. Note that the left and right arrows move the cursor and do not scroll the graphical display, as the Bode plotter always uses the horizontal and vertical settings that you specify to fit the data into a single screen. You can also use the mouse to drag the cursor or right-click the cursor to open a right-click menu similar to that for the oscilloscope cursor.

2) **Mode settings**

These settings determine the Bode plotter mode of operation. "Magnitude" will display the magnitude (gain) of the circuit frequency response. "Phase" will display the phase of the circuit frequency response.

3) **Horizontal settings**

These settings configure the horizontal axis for the Bode plotter display. "Log" specifies a logarithmic frequency scale and "Lin" specifies a linear frequency scale. "F" sets the final frequency and "I" sets the initial frequency for the measured frequency range.

4) **Vertical settings**

These settings configure the vertical axis for the Bode plotter display. "Log" specifies a logarithmic scale and "Lin" specifies a linear scale. "F" sets the final vertical axis value and "I" sets the initial vertical axis value.

If you select "Magnitude" as the Bode plotter mode then the vertical axis will represent gain (that is, the ratio of *Vout* to *Vin*). A linear scale will represent the magnitude as the ratio of *Vout* / *Vin*. A logarithmic scale will represent the magnitude in decibels.

If you select "Phase" as the Bode plotter mode then the vertical axis will be linear and have units of degrees.

5) **Plotter controls**

These settings allow you to work with the plotter data.

"Reverse" allows you to select between a white or black background for the Bode plotter display, just as for the two-channel oscilloscope display.

"Save" allows you to save the measured data to either a .BOD text file or a .TDM binary file.

"Set" allows you to determine the resolution of the displayed data (i.e., the total number of data points that the Bode plotter will collect).

34.3.3   Using the Bode Plotter

1.   Open the circuit file Ex34-02.

2.   Select the Bode plotter tool from the instrument bar and place it above the circuit.

3.   Connect the Bode plotter input across *VS* by connecting IN "+" between *VS* and *C1* and IN "–" to ground. The Bode plotter will use *VS* as the reference for the magnitude and phase measurements.

4.   Connect the Bode plotter output across *R1* by connecting OUT "+" between *C1* and *R1* and OUT "–" to ground. The Bode plotter will measure the magnitude and phase across *R1* relative to the input (*VS*).

5.   Double-click the Bode plotter to open the enlarged view.

6.   Set the Bode plotter settings to those in Table 34-3.

Table 34-3: Bode Plotter Settings

Mode	Horizontal	Vertical
Magnitude	Log F: 1 MHz I: 1 Hz	Log F: 20 dB I: −40 dB
Phase	Log F: 1 MHz I: 1 Hz	Lin F: 90 deg I: 0 deg

7.   Set the Mode setting to "Magnitude".

8.   Click the **Run** switch to simulate the circuit and wait for the Bode plotter to display the plot.

9.   Click the **Run** switch to stop the simulation. The display should look similar to Figure 34-5. As you should see, the magnitude of the output increases as the frequency increases and *C1* appears more and more as a short. Above a certain frequency, called the **corner frequency**, the magnitude levels out.

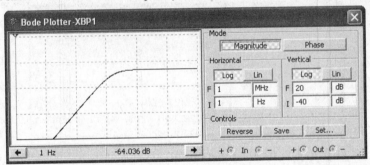

Figure 34-5: Example Bode Plot Magnitude Display

10.   Right-click the magnitude cursor to open the right-click menu and select "Set Y_Value =>".

11.   Enter "−3" for the value. The display should now look similar to Figure 34-6. As the display status bar shows, the −3 dB frequency is 1.596 Hz. You will learn more about −3 dB frequencies later.

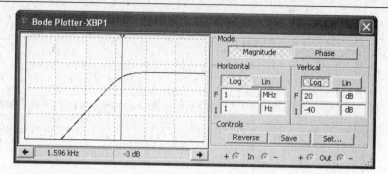

Figure 34-6: Bode Plotter with Repositioned Magnitude Cursor

12. Change the Mode setting to "Phase". The display should look similar to Figure 34-7. As you should see, the phase of the RC circuit changes from 90° at low frequency to 0° at high frequency.

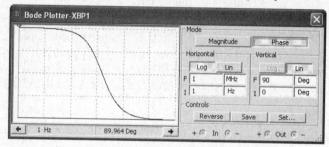

Figure 34-7: Example Bode Plot Phase Display

13. Right-click the phase cursor to open the right-click menu and select "Set Y_Value =>".

14. Enter "45" for the value. The display should now look similar to Figure 34-8. As the display status bar shows, the 45° frequency is 1.592 Hz.

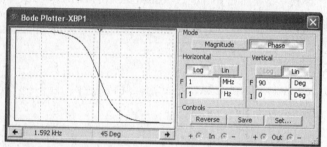

Figure 34-8: Bode Plotter with Repositioned Phase Cursor

15. Close the circuit file.

34.3.4   Bode Plotter Resonant Frequency Measurements

1. Open the circuit file Ex34-01.

2. Connect the Bode plot input across **VS** and the output across **R1**.

3. Set the values of **C1**, **L1**, and **R1** to each of the values shown in Table 34-4.

4. For each set of values, use the Bode plotter magnitude mode and cursor to measure the resonant frequency value f_r. You can do this by finding the maximum Y value of the plot with "Go to next Y_MAX =>" or "Go to next Y_MAX <=".

5. For each set of values, use the Bode plotter phase mode and cursor to measure and record the phase φ_r at resonance. You can do this by using the "Set X_Value" setting and specifying the resonant frequency you measured in Step 3.

Table 34-4: Bode Plotter Resonant Frequency Measurements

Circuit Value	$C1 = 510$ nF $L1 = 500$ µH $R1 = 1$ kΩ	$C1 = 220$ nF $L1 = 4.7$ mH $R1 = 2$ kΩ	$C1 = 360$ nF $L1 = 420$ µH $R1 = 510$ Ω	$C1 = 910$ nF $L1 = 27$ mH $R1 = 1.5$ kΩ
f_r				
φ_r				

Do your measured resonant frequency values in Table 34-4 agree with those in Table 34-2?

Is $\varphi_r \approx 0°$?

34.4 Application Exercise

A figure of merit that is often associated with resonant circuits is the Q, or quality, of the circuit. This value for series resonant circuits describes how quickly the series current approaches its maximum value as the circuit frequency approaches the resonant frequency f_r. The equation to find the Q value is

$$Q = f_r / BW$$
$$= f_r / (f_{CH} - f_{CL})$$

where

Q is the quality of the circuit,

f_r is the resonant frequency of the circuit,

BW is the bandwidth of the circuit, equal to $f_{CH} - f_{CL}$,

f_{CL} is the low −3 dB (low corner) frequency of the circuit, and

f_{CH} is the high −3 dB (high corner) frequency of the circuit.

Refer to Figure 34-9.

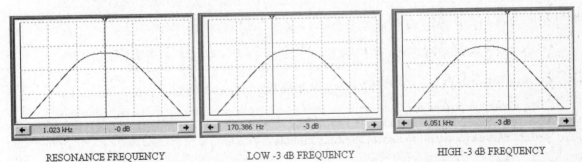

RESONANCE FREQUENCY   LOW -3 dB FREQUENCY   HIGH -3 dB FREQUENCY

Figure 34-9: Series Resonant Circuit Magnitude Frequency Response

For the frequency response shown, $f_r = 1023$ Hz, $f_{CL} = 170$ Hz, and $f_{CH} = 6051$ Hz so

$$BW = f_{CH} - f_{CL}$$
$$= 6051 \text{ Hz} - 170 \text{ Hz}$$
$$= 5881 \text{ Hz}$$
$$Q = f_r / BW$$

= 1023 Hz / 5881 Hz

= 0.174

1. Open the circuit file Pr34-01.

2. Connect the Bode plotter input across **VS** and the output across **R1**.

3. Set the Bode plotter settings to the values shown in Table 34-5.

Table 34-5: Bode Plotter Settings for *Q* Measurements

Mode	Horizontal	Vertical
Magnitude	Log F: 1 MHz I: 1 Hz	Log F: 20 dB I: −40 dB

4. Change the value of **R1** to 1 kΩ.

5. Click the **Run** switch to simulate the circuit and wait for the Bode plotter to display the plot.

6. Click the **Run** switch to stop the circuit simulation.

7. Measure and record the resonant frequency, -3 dB frequencies, bandwidth, and *Q* values for **R1** = 1 kΩ in Table 34-6.

8. Repeat Steps 4 through 6 for **R1** = 5 kΩ and **R1** = 10 kΩ.

Table 34-6: Series Resonant Circuit *Q* Measurements

Circuit Value	*R1* = 1 kΩ	*R1* = 5 kΩ	*R1* = 10 kΩ
f_r			
f_{CL}			
f_{CH}			
BW = $f_{CH} - f_{CL}$			
$Q = f_r$ / BW			

How does a larger value of **R1** affect the value of f_r for a series resonant circuit?

How does a larger value of **R1** affect the value of *Q* for a series resonant circuit?

34.5 Section Summary

This section introduced the Multisim Bode plotter and explored some of the characteristics of series resonant circuits. As the frequency of a series RLC circuit approaches the resonant frequency, the inductive and capacitive reactances approach the same value until at resonance they cancel. At this point the inductive and capacitive voltages cancel, circuit impedance is equal to the circuit resistance, and the current in the circuit reaches its maximum value. The value of series resistance in the circuit will not affect the resonant frequency of the circuit, but will affect the *Q* of the circuit.

35. Parallel Resonance

35.1 Introduction

In a series RLC circuit at resonance the voltages across the capacitor and inductor are equal and opposite and cancel. In a parallel RLC circuit at resonance the currents through the inductor and capacitor are equal and opposite and cancel. To see why, refer to the series-parallel RLC circuit of Figure 35-1.

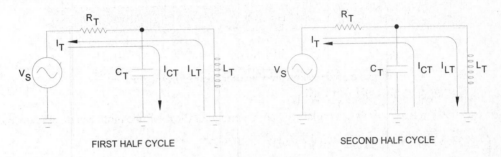

Figure 35-1: Series-Parallel RLC Circuit

Because C_T and L_T are in parallel the voltage across them is the same. The capacitor current leads the voltage by 90° and the inductor current lags the voltage by 90°, so I_{CT} and I_{LT} are 180° out of phase (i.e., have opposite polarity) as Figure 35-1 shows. I_T is therefore the difference between I_{CT} and I_{LT}. At the resonant frequency X_{CT} and X_{LT} are equal, so that I_{CT} and I_{LT} are also equal and the value of $I_T = 0$. When this occurs the only currents present are through C_T and L_T, as shown in Figure 35-2.

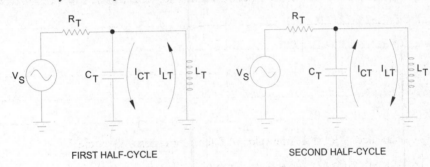

Figure 35-2: Parallel LC Currents at Resonance

On one half-cycle the inductor de-energizes and its current charges the capacitor. When the inductor is fully de-energized and the capacitor is fully charged the process reverses, so that on the next half-cycle the capacitor discharges and its current energizes the inductor. When the capacitor is fully discharged and the inductor is fully energized the process again reverses and the cycle repeats. This alternating current develops the ac voltage across the parallel combination of L_T and C_T even though in an ideal circuit no current flows from V_S. The parallel combination of L_T and C_T is often referred to as a **tank circuit** or **tuned circuit.** The alternating current within the tank circuit is sometimes referred to as the **tank current** or **slosh current**, as the current resembles water sloshing back and forth inside a tank. Oscillator designs often use LC tank circuits to set the oscillator frequency.

In this section you will:

- Verify the resonant frequency for parallel RLC circuits.

- Investigate the behavior of parallel RLC circuits at resonance.

- Use the Multisim software to verify your analysis of series RLC circuits at resonance.

35.2 Pre-Lab

Refer to the parallel RLC circuit in Figure 35-3.

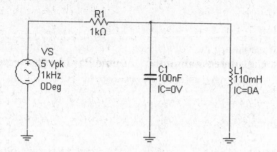

Figure 35-3: Example Parallel Resonant Circuit

The resonant frequency f_r for this circuit is

$$f_r = \frac{1}{2\pi\sqrt{(L1)(C1)}} = \frac{1}{2\pi\sqrt{(100\text{ nF})(110\text{ mH})}} = 1.52\text{ kHz}$$

At this frequency

$$X_{C1} = -j\,(1\,/\,[2\pi(f_r)(C1)])$$

$$= -j(1\,/\,[(2\pi)(1.52\text{ kHz})(100\text{ nF})])$$

$$= -j(1.05\text{ k}\Omega) = 1.05\text{ k}\Omega \angle -90°$$

$$X_{L1} = j[2\pi(f_r)(L1)]$$

$$= j[(2\pi)(1.52\text{ kHz})(110\text{ mH})]$$

$$= j(1.05\text{ k}\Omega) = 1.05\text{ k}\Omega \angle 90°$$

So

$$I_{C1} = VS\,/\,X_{L1}$$

$$= (5.0\text{ Vpk} \angle 0°)\,/\,(1.05\text{ k}\Omega \angle -90°)$$

$$= 4.76\text{ mApk} \angle 90° = j(4.76\text{ mApk})$$

$$= 3.37\text{ mA}_{RMS} = j(3.37\text{ mA}_{RMS})$$

$$I_{L1} = VS\,/\,X_{C1}$$

$$= (5.0\text{ Vpk} \angle 0°)\,/\,(1.05\text{ k}\Omega \angle 90°)$$

$$= 4.76\text{ mApk} \angle -90° = -j(4.76\text{ mApk})$$

$$= 3.37\text{ mA}_{RMS} \angle -90° = -j(3.37\text{ mA}_{RMS})$$

$$I_T = I_{XC1} + I_{XL1}$$

$$= j(3.37\text{ mA}_{RMS}) - j(3.37\text{ mA}_{RMS})$$

$$= j(3.37\text{ mA}_{RMS} - 3.37\text{ mA}_{RMS})$$

$$= j(0\text{ mA}_{RMS}) = 0\text{ mA}_{RMS} \angle 0°$$

The voltage across $L1$ and $C1$ is then

$$V_{C1} = V_{L1} = VS - (I_T \times R1)$$

$$= (5.0 \text{ Vpk} \angle 0°) - (0 \text{ mA}_{RMS} \angle 90°) (1 \text{ k}\Omega \angle 0°)$$

$$= (3.54 \text{ V}_{RMS} \angle 0°) - (0 \text{ mA}_{RMS} \angle 90°)(1 \text{ k}\Omega \angle 0°)$$

$$= 3.54 \text{ V}_{RMS} - 0 \text{ Vpk} = 3.54 \text{ V}_{RMS}$$

Calculate the resonant frequencies and circuit values at resonance for each set of values in Table 35-1.

Table 35-1: Calculated Values for Example Parallel Resonant Circuit

Circuit Value	$C1 = 100$ nF $L1 = 110$ mH $R1 = 1$ kΩ	$C1 = 220$ nF $L1 = 130$ mH $R1 = 1.5$ kΩ	$C1 = 1.0$ μF $L1 = 1.0$ mH $R1 = 200$ Ω	$C1 = 2.2$ μF $L1 = 120$ μH $R1 = 10$ Ω
f_r	1.52 kHz			
X_{C1}	1.05 kΩ $\angle -90°$			
X_{L1}	1.05 kΩ $\angle 90°$			
$I_{C1} = 0.707 \times (VS / X_{C1})$	3.37 mA$_{RMS}$ $\angle 90°$			
$I_{L1} = 0.707 (VS / I_{L1})$	3.37 mA$_{RMS}$ $\angle -90°$			
$I_T = I_{C1} + I_{L1}$	0 mA$_{RMS}$ $\angle 0°$			
$V_{R1} = I_T \times R1$	0 V$_{RMS}$ $\angle 90°$			
$V_{L1} = V_{C1} = VS - V_{R1}$	3.54 V$_{RMS}$ $\angle 0°$			

35.3 Design Verification

Open the circuit file Ex35-01. Set the values of *C1*, *L1*, and *R1* to each set of values shown in Table 35-2 and set the frequency of the ac voltage source to each of the corresponding resonant frequencies you calculated in Table 35-1. Use a multimeter to measure I_{C1}, I_{L1}, and I_{R1}, and record the values in the appropriate column of Table 35-2.

Table 35-2: Measured Values for Example Parallel Resonant Circuit

Circuit Value	$C1 = 100$ nF $L1 = 110$ mH $R1 = 1$ kΩ	$C1 = 220$ nF $L1 = 130$ mH $R1 = 1.5$ kΩ	$C1 = 1.0$ μF $L1 = 1.0$ mH $R1 = 200$ Ω	$C1 = 2.2$ μF $L1 = 120$ μH $R1 = 10$ Ω
f_r	1.52 kHz			
I_{C1}				
I_{L1}				
I_{R1}				

Do your measured current values in Table 35-2 agree with your calculated values in Table 35-1?

Connect the Bode plotter input across *VS* and the output across the parallel combination of *C1* and *L1*. Set the values of *C1*, *L1*, and *R1* to each set of values shown in Table 35-3. Use the Bode plotter to determine the resonant frequency for each set of values and record the values in Table 35-3.

Table 35-3: Bode Plotter Resonant Frequency Measurements

Circuit Value	$C1 = 100$ nF $L1 = 110$ mH $R1 = 1$ kΩ	$C1 = 220$ nF $L1 = 130$ mH $R1 = 1.5$ kΩ	$C1 = 1.0$ μF $L1 = 1.0$ mH $R1 = 200$ Ω	$C1 = 2.2$ μF $L1 = 120$ μH $R1 = 10$ Ω
f_r				

Do your measured values in Table 35-3 agree with your calculated values in Table 35-1?

35.4 Application Exercise

1. Open the circuit file Pr35-01.

2. Connect the Bode plotter input across *VS* and the output across the parallel combination of *C1* and *L1*.

3. Set the Bode plotter settings to the values shown in Table 35-4.

Table 35-4: Bode Plotter Settings for *Q* Measurements

Mode	Horizontal	Vertical
Magnitude	Log F: 1 MHz I: 1 Hz	Log F: 20 dB I: −40 dB

4. Change the value of *R1* to 100 Ω.

5. Click the **Run** switch to simulate the circuit and wait for the Bode plotter to display the plot.

6. Click the **Run** switch to stop the circuit simulation.

7. Measure the resonant frequency, −3 dB frequencies, bandwidth, and *Q* values for *R1* = 1 kΩ and record the measured values in Table 35-5.

8. Repeat Steps 4 through 7 for *R1* = 500 Ω and *R1* = 1 kΩ.

Table 35-5: Parallel Resonant Circuit *Q* Measurements

Circuit Value	*R1* = 100 Ω	*R1* = 500 Ω	*R1* = 1 kΩ
f_r			
f_{CL}			
f_{CH}			
BW = $f_{CH} - f_{CL}$			
$Q = f_r / $ BW			

How does a larger value of *R1* affect the value of f_r for a parallel resonant circuit?

How does a larger value of *R1* affect the value of *Q* for a parallel resonant circuit?

35.5 Section Summary

This section explored some of the characteristics of parallel resonant circuits. As the frequency of a parallel RLC circuit approaches the resonant frequency, the inductive and capacitive reactances approach the same value until at resonance they cancel. At this point the inductive and capacitive currents cancel, circuit impedance is infinite, and the total current in the circuit reaches its minimum value. At resonance the parallel combination of capacitance and inductance form a tank circuit. This circuit contains an ac current that transfers energy between the magnetic field of the inductor and electrostatic field of the capacitor at the resonant frequency. The value of series resistance in the circuit will not affect the resonant frequency of the circuit, but will affect the *Q* of the circuit.

36. Transformer Circuits

36.1 Introduction

Transformers are magnetic devices whose basic operation is similar to that of inductors. Just as for inductors, a changing current through a transformer coil will induce a voltage across the coil and create a changing electromagnetic field. A difference is that the design of a basic transformer applies, or couples, the changing electromagnetic field of the primary coil to a secondary coil, and this coupled field induces a voltage across the secondary coil. An ideal transformer will couple all the electromagnetic flux from the primary coil to the secondary coil(s) so that it transfers 100% of the power from the primary side to the secondary side. In practice the physical construction and core material determine the efficiency of the transformer, but the operation of most practical transformers is very close to the ideal.

Transformers have several applications. One application is to provide electrical isolation. Because transformers use an electromagnetic field to transfer energy there is no dc path between the transformer primary and secondary. Another application is to change the amplitude of an ac voltage. A transformer that increases the amplitude is called a step-up transformer, and a transformer that decreases the amplitude is called a step-down transformer. A third application is impedance matching. Maximum power transfer requires that the source resistance equal the load resistance. When this is not the case you can sometimes use a transformer to match the impedances.

In this section you will:

- Study the operation of step-up and step-down transformers.

- Investigate the results of using a transformer for impedance matching.

- Use the Multisim software to verify the voltage, current, and power characteristics of transformers.

36.2 Pre-Lab

One equation that relates the primary voltage V_{PRI}, secondary voltage V_{SEC}, and turns ratio n is

$$V_{SEC} = V_{PRI} \times n \rightarrow n = V_{SEC} / V_{PRI}$$

Another is

$$V_{PRI} = V_{SEC} \times n \rightarrow n = V_{PRI} / V_{SEC}$$

The Multisim program uses the latter. When you must determine the proper turns ratio for a transformer circuit, divide the applied primary voltage value by the desired secondary voltage.

36.2.1 Step-Down Transformer Circuits

Refer to Figure 36-1.

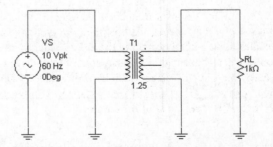

Figure 36-1: Example Step-Down Transformer Circuit

For each of the turns ratio values in Table 36-1, calculate the peak ac voltage across *R1*.

Table 36-1: Calculated Secondary Voltages for Example Step-Down Transformer Circuit

n	$V_{SEC} = 10 \text{ Vpk} / n$
1.25	
1.50	
2.50	
3.33	
6.25	

For each of the peak secondary voltages in Table 36-2, calculate the necessary transformer turns ratio.

Table 36-2: Calculated Turns Ratio for Example Step-Down Transformer Circuit

V_{SEC} (Vpk)	$n = 10 \text{ Vpk} / Vs$
0.75	
1.25	
1.5	
3.3	
6.0	

36.2.2 Step-Up Transformer Circuit

Refer to Figure 36-2.

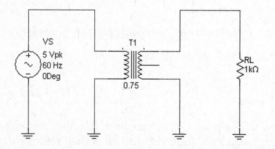

Figure 36-2: Example Step-Up Transformer Circuit

For each of the turns ratio values in Table 36-3, calculate the peak ac voltage across **R1**.

Table 36-3: Calculated Secondary Voltages for Example Step-Up Transformer Circuit

n	$V_{SEC} = 10 \text{ Vpk} / n$
0.7500	
0.6666	
0.6250	
0.3333	
0.1250	

For each of the peak secondary voltages in Table 36-4, calculate the necessary transformer turns ratio.

Table 36-4: Calculated Turns Ratio for Example Step-Up Transformer Circuit

V_{SEC} (Vpk)	$n = 10$ Vpk / V_S
12.5	
15.0	
20.0	
24.0	
48.0	

36.2.3 Impedance Matching Transformer Circuit

A transformer circuit transfers maximum power when the reflected load in the primary equals the source resistance. The load $R_L = V_{SEC} / I_{SEC}$ and the reflected load $R_{PRI} = V_{PRI} / I_{PRI}$, so

$$R_{PRI} / R_L = (V_{PRI} / I_{PRI}) / (V_{SEC} / I_{SEC})$$

$$= (V_{PRI} / I_{PRI}) (I_{SEC} / V_{SEC})$$

$$= (V_{PRI} / V_{SEC}) (I_{SEC} / I_{PRI})$$

Using the Multisim program convention for turns ratio $n = V_{PRI} / V_{SEC} = I_{SEC} / I_{PRI}$, this gives

$$R_{PRI} / R_L = (n) (n)$$

$$= n^2$$

So

$$R_{PRI} = (n^2) R_L$$

Refer to Figure 36-3.

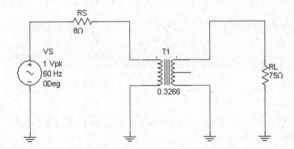

Figure 36-3: Example Impedance Matching Transformer Circuit

The circuit transfers maximum power from the source voltage when $R_{PRI} = R_S$, so

$$R_S = R_{PRI}$$

$$= (n^2) R_L$$

From this $n^2 = R_S / R_L$, so

$$n = \sqrt{R_S / R_L}$$

For the circuit values in Figure 36-3, the circuit will transfer maximum power for $n = \sqrt{(8\,\Omega) / (75\,\Omega)} = 0.3266$, as shown.

For each of the load resistance values in Table 36-5, calculate the load voltage and load power without impedance matching.

Table 36-5: Calculated Values for Circuit of Figure 36-3 Without Impedance Matching

	$RL = 10\ \Omega$	$RL = 25\ \Omega$	$RL = 50\ \Omega$	$RL = 100\ \Omega$
$V_{RL} = VS \times [RL / (RS + RL)]$				
$P_{RL} = (0.707 \times V_{RL})^2 / RL$				

For each of the load resistances in Table 36-6, calculate the turns ratio required for maximum power transfer, reflected load, primary voltage, load voltage, and load power with impedance matching.

Table 36-6: Calculated Values for Circuit of Figure 36-3 With Impedance Matching

	$RL = 10\ \Omega$	$RL = 25\ \Omega$	$RL = 50\ \Omega$	$RL = 100\ \Omega$
$n = \sqrt{8\ \Omega / RL}$				
$R_{PRI} = (n^2)\ RL$				
$V_{PRI} = VS \times [R_{PRI} / (RS + R_{PRI})]$				
$V_{RL} = V_{PRI} / n$				
$P_{RL} = (0.707 \times V_{RL})^2 / RL$				

36.3 Design Verification

36.3.1 Step-Down Transformer Circuit

Open the circuit file Ex36-01. Double-click the transformer to open the TS_Virtual properties window and change the "Primary-to-Secondary Turns Ratio" to the indicated turns ratios shown in Table 36-7. For each value, use the oscilloscope to measure and record the peak secondary voltage across *RL*.

Table 36-7: Measured Secondary Voltages for Example Step-Down Transformer Circuit

n	Measured V_{SEC} (Vpk)
1.25	
1.50	
2.50	
3.33	
6.25	

Do your measured values in Table 36-7 agree with your calculated values in Table 36-1?

Open the TS_Virtual properties window and change the "Primary-to-Secondary Turns Ratio" to your calculated turns ratios in Table 36-2 for the secondary voltages shown in Table 36-8. For each value, use the oscilloscope to measure and record the peak secondary voltage across *RL*.

Table 36-8: Measured Secondary Voltages for Calculated Turns Ratio for Example Step-Down Transformer Circuit

Target V_{SEC} (Vpk)	Calculated n	Measured V_{SEC} (Vpk)
0.75		
1.25		
1.5		
3.3		
6.0		

Do your measured secondary voltages in Table 36-8 agree with the target values in Table 36-2?

36.3.2 Step-Up Transformer Circuit

Open the circuit file Ex36-02. Double-click the transformer to open the TS_Virtual properties window and change the "Primary-to-Secondary Turns Ratio" to the indicated turns ratios shown in Table 36-9. For each value, use the oscilloscope to measure and record the peak secondary voltage across **RL**.

Table 36-9: Measured Secondary Voltages for Example Step-Up Transformer Circuit

n	$V_{SEC} = 10$ Vpk / n
0.7500	
0.6666	
0.6250	
0.3333	
0.1250	

Do your measured values in Table 36-9 agree with your calculated values in Table 36-3?

Open the TS_Virtual properties window and change the "Primary-to-Secondary Turns Ratio" to your calculated turns ratios in Table 36-4 for each of the secondary voltages shown in Table 36-10. For each value, use the oscilloscope to measure and record the peak secondary voltage across **RL**.

Table 36-10: Measured Secondary Voltages for Calculated Turns Ratio for Example Step-Up Transformer Circuit

Target V_{SEC} (Vpk)	Calculated n	Measured V_{SEC} (Vpk)
12.5		
15.0		
20.0		
24.0		
48.0		

Do your measured secondary voltages in Table 36-10 agree with the target values in Table 36-4?

36.3.3 Impedance Matching Transformer Circuit

Open the circuit file Ex36-03. Double-click the transformer to open the TS_Virtual properties window and change the "Primary-to-Secondary Turns Ratio" to 1 to eliminate the effects of the transformer. For each value in Table 36-11, use the oscilloscope to measure and record the peak secondary voltage across **RL**. Calculate and record the load power P_{RL} for each.

Table 36-11: Measured Values for Circuit of Figure 36-3 Without Impedance Matching

Circuit Value	$RL = 10\ \Omega$	$RL = 25\ \Omega$	$RL = 50\ \Omega$	$RL = 100\ \Omega$
Measured V_{RL} (Vpk)				
$P_{RL} = (0.707 \times V_{RL})^2 / RL$				

Do your measured values for load voltage and power in Table 36-11 agree with your calculated values in Table 36-5?

Open the TS_Virtual properties window and change the "Primary-to-Secondary Turns Ratio" to your calculated turns ratios in Table 36-6 for each of the load resistances shown in Table 36-12. For each value, use the oscilloscope to measure and record the peak secondary voltage across **RL**. Calculate and record the load power P_{RL} for each.

Table 36-12: Measured Values for Circuit of Figure 36-3 With Impedance Matching

Circuit Value	$RL = 10\ \Omega$	$RL = 25\ \Omega$	$RL = 50\ \Omega$	$RL = 100\ \Omega$
Calculated n				
Measured V_{RL} (peak)				
$P_{RL} = (0.707 \times V_{RL})^2 / RL$				

Do your measured values for load voltage and power in Table 36-12 agree with your calculated values in Table 36-6?

36.4 Application Exercise

A common misconception is that transformers can multiply or increase power. Transformers change ac voltage and current values from a source to a load, but cannot provide more power from the secondary to a load than the source provides to the primary. An ideal transformer transfers all the power from the primary to the secondary, but practical transformers always have some loss. The efficiency η (eta) of a transformer indicates how much power the transformer transfers from the primary to the secondary and is given by

$$\eta = (P_{OUT} / P_{IN}) \times 100\%$$

100% efficiency means that $P_{OUT} = P_{IN}$ and that the transformer passes all the input power to the output. 50% efficiency means that $P_{OUT} = P_{IN}/2$ and that the transformer passes only half the input power to the output.

Open the circuit file Pr36-01. Change the turns ratio of the transformer to each of the values in Table 36-13. Measure and record the input and output power indicated by the wattmeters and calculate the efficiency for each turns ratio value.

Table 36-13: Measured Input and Output Powers

n	P_{IN}	P_{OUT}	$\eta = P_{OUT} / P_{IN}$
0.1			
0.2			
0.5			
1			
2			
5			
10			
20			
50			
100			

Is $P_{IN} \approx P_{OUT}$ for each value of n?

Is $\eta > 100\%$ for any value of n?

36.5 Section Summary

Transformers are magnetic devices that can transfer energy while maintaining dc isolation. This section covered the operation and characteristics of transformer step-up, step-down, and impedance matching applications. Circuits can use transformers to increase or decrease the amplitudes of ac currents and voltages from an ac source, but cannot provide more power to the load than the voltage supply provides to the transformer.

37. Low-Pass Filter Circuits

37.1 Introduction

37.1.1 Low-Pass Filter Basics

Filters are circuits that selectively pass or reject (block) electrical signals on the basis of frequency. Typical filter applications are isolating signals of interest and removing electrical noise or other unwanted signals. This section will investigate the behavior of passive low-pass filters. A passive filter is one that consists solely of passive components, such as resistors, capacitors, and inductors.

There are two basic methods to implement a passive low-pass filter. The first way is to shunt the high frequencies to ground. The second way is to block the high frequencies from reaching the output. Figure 37-1 shows these methods.

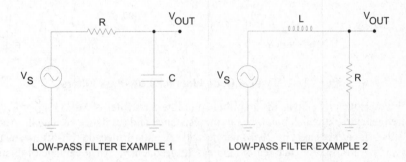

LOW-PASS FILTER EXAMPLE 1   LOW-PASS FILTER EXAMPLE 2

Figure 37-1: Examples of Passive Low-Pass Filters

In the first example, the capacitor acts as an open at low frequencies so that low frequencies pass through R to the output. At high frequencies the capacitor acts as a short so that high frequencies are shorted to ground.

In the second example the inductor acts as a short at low frequencies so that low frequencies pass through to the output. At high frequencies the inductor acts as an open so that high frequencies do not reach the output and V_{OUT} is pulled to ground through R.

In this section you will:

- Learn about the use of Bode plots for representing the frequency response of passive low-pass filters.

- Study the amplitude and phase response of passive low-pass filters.

- Investigate the effects of loading on passive low-pass filter response.

- Use the Multisim software to verify the characteristics of passive low-pass filters.

37.1.2 Bode Plots

Bode plots are typically used to represent the frequency response of a filter. A Bode plot is a graph that shows the amplitude or phase of a filter as a function of frequency. A typical Bode plot represents the amplitude A as a ratio of the output voltage to the input voltage in decibels. For $A = V_{OUT} / V_{IN}$

$$A \text{ (dB)} = 20 \log (A) = 20 \log (V_{OUT} / V_{IN})$$

If the input and output are equal, then $A = V_{OUT} / V_{IN} = 1$ and A (dB) = 20 log (1) dB = (20)(0) dB = 0 dB. If $A = V_{OUT} / V_{IN} = 10$, then A (dB) = 20 log (10) dB = (20)(1) dB = 20 dB. Similarly, if $A = V_{OUT} / V_{IN} = 0.1$ then A (dB) = 20 log (0.1) dB = (20) (−1) dB = −20 dB. Decibels allow you to represent a very large range of values in a compact form, as each increase or decrease of 20 dB represents a tenfold increase or decrease in amplitude.

A typical Bode plot also uses a logarithmic scale to show frequency. Each tenfold increase or decrease in frequency is called a "decade", while each twofold increase or decrease in frequency is called an "octave".

Refer to the Bode plots in Figure 37-2. These plots are for the second (RL) example of a low-pass filter shown in Figure 37-1.

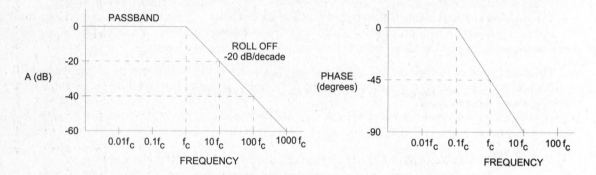

Figure 37-2: Typical Bode Plots for Low-Pass Filters

As the amplitude Bode plot shows, the amplitude of a low-pass filter is 0 dB ($V_{OUT} = V_{IN}$) until it reaches the corner frequency f_C, at which point the circuit resistance and reactance are equal. Above f_C the amplitude rolls off, or decreases, by 20 dB per decade. A −20 dB/decade roll-off means that the amplitude decreases by a factor of ten for each tenfold increase in frequency. An advantage of this is that the Bode plot linearizes the roll-off so that you can easily interpolate values for any frequency above f_C. At a frequency f that is N decades above f_C (where $N = \log (f / f_C)$) the amplitude $A = N \times -20$ dB.

Although the Bode plot shows that the amplitude is 0 dB up to the corner frequency and starts to roll off exactly at f_C, this is part of the idealized plot. At the corner frequency, $V_{OUT} / V_{IN} = 0.707$ so that the amplitude is −3 dB. At frequencies near f_C the amplitude is actually in a transition region. A rule of thumb is that frequencies below $f_C / 10$ are in the filter passband. The passband is so named because the filter passes signals in this band of frequencies to the output so that $V_{OUT} = V_{IN}$.

The phase Bode plot is similar to the amplitude Bode plot, but the units are in degrees. For frequencies below $f_C / 10$ (in the passband) the phase is considered to be 0°, and above $10 f_C$ the phase is considered to be −90°. Between $f_C / 10$ and $10 f_C$ the phase transitions from 0° to 90° with a phase of −45° at the corner frequency.

37.2 Pre-Lab

The corner frequency of a low-pass filter is the frequency at which the resistance and reactance are equal. For this to be true for the first low-pass filter example in Figure 37-1

$$R = X_C = 1 / (2\pi f_C C)$$

so

$$f_C = 1 / (2\pi RC)$$

For the second low-pass filter example in Figure 37-1

$$R = X_L = 2\pi f_C L$$

so

$$f_C = R / (2\pi L)$$

37.2.1 Low-Pass Filter Circuit Example 1

Refer to the circuit in Figure 37-3.

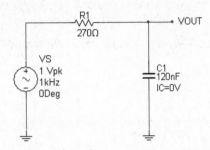

Figure 37-3: Low-Pass Filter Circuit Example 1

Calculate the circuit values indicated in Table 37-1 for the frequencies indicated. Recall for this type of low-pass filter that $f_C = 1 / [2\pi(R1)(C1)]$.

Table 37-1: Calculated Circuit Values for Low-Pass Filter Circuit Example 1

Circuit Value	$f = f_C / 10$	$f = f_C$	$f = 10 f_C$
f			
$X_{C1} = 1 / [2\pi f (C1)]$			
$Z_T = \sqrt{R1^2 + X_{C1}^2}$			
$VOUT = VS \times (X_{C1} / Z_T)$			
$VOUT \text{ (dB)} = -20 \log (VOUT)$			
$\varphi = -\tan^{-1}(R1 / X_{C1})$			

Calculate the decades N above f_C and the amplitude A in decibels for each of the frequencies f shown in Table 37-2.

Table 37-2: Calculated Amplitude Values for Low-Pass Filter Circuit Example 1

f	$N = \log (f / f_C)$	$A = N \times -20 \text{ dB}$
50 kHz		
100 kHz		
200 kHz		
500 kHz		
1 MHz		

37.2.2 Low-Pass Filter Circuit Example 2

Refer to the circuit in Figure 37-4.

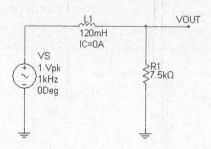

Figure 37-4: Low-Pass Filter Circuit Example 2

Calculate the circuit values indicated in Table 37-3 for the frequencies indicated. Recall for this type of low-pass filter that $f_C = R1 / [2\pi(L1)]$.

Table 37-3: Calculated Circuit Values for Low-Pass Filter Circuit Example 2

Circuit Value	$f = f_C / 10$	$f = f_C$	$f = 10 f_C$
f			
$X_{L1} = 2\pi f(L1)$			
$Z_T = \sqrt{R1^2 + X_{L1}^2}$			
$VOUT = VS \times (X_{L1} / Z_T)$			
$VOUT$ (dB) $= -20 \log (VOUT)$			
$\varphi = -\tan^{-1}(X_{L1} / R1)$			

Calculate the decades N above f_C and the amplitude A in decibels for each of the frequencies f shown in Table 37-4.

Table 37-4: Calculated Amplitude Values for Low-Pass Filter Circuit Example 2

f	$N = \log (f / f_C)$	$A = N \times -20$ dB
100 kHz		
200 kHz		
500 kHz		
1 MHz		
2 MHz		

37.3 Design Verification

37.3.1 Low-Pass Filter Circuit Example 1

1. Open the circuit file Ex37-01.

2. Connect the input of the Bode plotter across **VS** and the output between **VOUT** and ground.

3. Set the Bode plotter settings to those in Table 37-5.

Table 37-5: Bode Plotter Settings

Mode	Horizontal	Vertical
Magnitude	Log F: 5 MHz I: 1 Hz	Log F: 20 dB I: −60 dB
Phase	Log F: 1 MHz I: 1 Hz	Lin F: 95 deg I: −95 deg

4. Measure the circuit amplitude and phase readings for each of the frequencies you calculated for Table 37-1 and record the values in Table 37-6.

Table 37-6: Measured Circuit Values for Low-Pass Filter Circuit Example 1

Circuit Value	$f = f_C / 10$	$f = f_C$	$f = 10 f_C$
f			
Magnitude (dB)			
Phase (degrees)			

Do the measured circuit values in Table 37-6 agree with your calculated values in Table 37-1?

5. Measure the amplitude A for each of the frequencies f shown in Table 37-7 and record the values.

Table 37-7: Measured Amplitude Values for Low-Pass Filter Circuit Example 1

f	A
50 kHz	
100 kHz	
200 kHz	
500 kHz	
1 MHz	

Do your measured circuit values in Table 37-7 agree with your calculated values in Table 37-2?

37.3.2 Low-Pass Filter Circuit Example 2

1. Open the circuit file Ex37-02.

2. Connect the input of the Bode plotter across **VS** and the output between **VOUT** and ground.

3. Set the Bode plotter settings to those in Table 37-8.

Table 37-8: Bode Plotter Settings

Mode	Horizontal	Vertical
Magnitude	Log F: 5 MHz I: 1 Hz	Log F: 20 dB I: −60 dB
Phase	Log F: 1 MHz I: 1 Hz	Lin F: 95 deg I: −95 deg

4. Measure the circuit amplitude and phase readings for each of the frequencies you calculated for Table 37-3 and record the values in Table 37-9.

Table 37-9: Measured Circuit Values for Low-Pass Filter Circuit Example 2

Circuit Value	$f = f_C / 10$	$f = f_C$	$f = 10 f_C$
f			
Magnitude (dB)			
Phase (degrees)			

Do your measured circuit values in Table 37-9 agree with your calculated values in Table 37-3?

5. Measure the amplitude A for each of the frequencies f shown in Table 37-10 and record the values.

Table 37-10: Measured Amplitude Values for Low-Pass Filter Circuit Example 2

f	A
100 kHz	
200 kHz	
500 kHz	
1 MHz	
2 MHz	

Do the measured circuit values in Table 37-10 agree with your calculated values in Table 37-4?

37.4 Application Exercise

Just as a load can affect the output of a voltage divider, a load can affect the frequency response of a filter circuit. Refer to the loaded filter circuits and their circuit equivalents in Figure 37-5 and Figure 37-6.

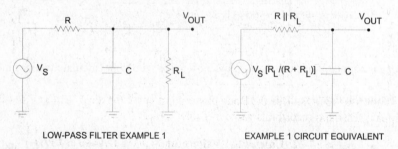

Figure 37-5: Loaded Filter Circuit Example 1 and Circuit Equivalent

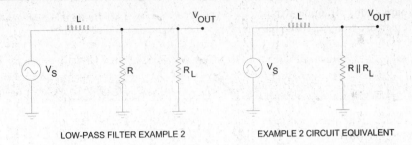

LOW-PASS FILTER EXAMPLE 2   EXAMPLE 2 CIRCUIT EQUIVALENT

Figure 37-6: Loaded Filter Circuit Example 2 and Circuit Equivalent

As you can see, the circuit equivalents for both loaded circuits have the same form as the original unloaded circuit but with different circuit values. The circuit equivalents for both loaded low-pass filters have a resistance value of $R \parallel R_L$ so that the corner frequency is affected. In addition, the circuit equivalent in Figure 37-5, which you can obtain by Thevenizing the circuit, shows that a load resistor will affect the amplitude of the output voltage as well.

For the circuit in Figure 37-7, calculate and record the equivalent resistance and corner frequency in Table 37-11.

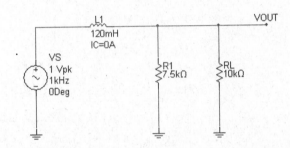

Figure 37-7: Low-Pass Filter Circuit with 10 kΩ Load

Table 37-11: Calculated Values for Loaded Low-Pass Filter

Circuit Value	Calculated Value
$R_{EQ} = R1 \parallel RL$	
$f_C = R_{EQ} / [(2\pi)(L1)]$	

Open the circuit file Pr37-01. Use the Bode plotter to measure the ideal −3 dB corner frequency for the loaded low-pass filter and record this value in Table 37-12.

Table 37-12: Calculated Values for Loaded Low-Pass Filter

Circuit Value	Measured Value
f_C	

How does your measured value of f_C in Table 37-12 compare with your measured value of f_C in Table 37-9?

37.5 Section Summary

This section presented the basic attributes of passive low-pass filters and the Bode plot used to describe the amplitude and phase frequency response of filters. Low-pass filters pass frequencies below a specific frequency, called the corner frequency. The low-pass filter blocks frequencies above this corner frequency so that the output rolls off, or decreases, 20 dB per decade. Low-pass filters also exhibit a phase shift as the frequencies increase from passband through the corner frequency and beyond.

38. High-Pass Filter Circuits

38.1 Introduction

38.1.1 High-Pass Filter Basics

High-pass filters are very similar to low-pass filters. You can implement a high-pass filter by exchanging the components in a low-pass filter as shown in Figure 38-1. In the first example the capacitor blocks the low frequencies from reaching the output. In the second example the inductor shunts the low frequencies to ground.

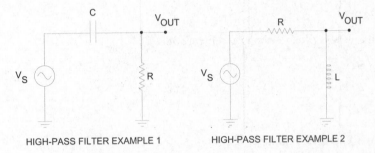

HIGH-PASS FILTER EXAMPLE 1   HIGH-PASS FILTER EXAMPLE 2

Figure 38-1: Examples of Passive High-Pass Filters

In this section you will:

- Study the amplitude and phase response of passive high-pass filters.

- Investigate the effects of loading on passive high-pass filter response.

- Use the Multisim software to verify the characteristics of passive high-pass filters.

38.1.2 Bode Plots

Refer to the Bode plots in Figure 38-2. These plots are for the first example of a high-pass filter shown in Figure 38-1.

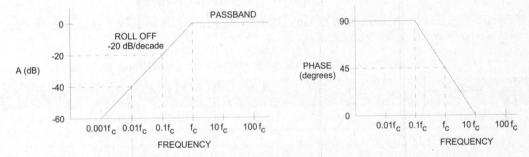

Figure 38-2: Typical Bode Plots for High-Pass Filters

The amplitude Bode plot shows that the amplitude of a high-pass filter increases 20 dB per decade until it reaches the corner frequency, after which the amplitude remains at 0 dB ($V_{OUT} = V_{IN}$). As for the low-pass filter the Bode plot linearizes the roll-off region. At a frequency f that is N decades below f_C (where $N = \log (f / f_C)$), the amplitude $A = N \times 20$ dB. Note that N will be a negative value for $f < f_C$.

Although the Bode plot shows that the amplitude is 0 dB at and above f_C, this is part of the idealized plot. At the corner frequency $V_{OUT} / V_{IN} = 0.707$ so that the amplitude is −3 dB. At frequencies near f_C the amplitude is actually in a transition region. A rule of thumb is that frequencies above $10\,f_C$ are in the filter passband.

The phase Bode plot for the high-pass filter is similar to that for a low-pass filter. For frequencies below $f_C / 10$ (in the passband) in the plot for Figure 38-1 the phase is considered to be 90° and above $10 f_C$ the phase is considered to be 0°. Between $f_C / 10$ and $10 f_C$ the phase transitions from 90° to 0° with a phase of 45° at the corner frequency.

38.2 Pre-Lab

The corner frequency of a high-pass filter is the frequency at which the resistance and reactance are equal. For this to be true for the first high-pass filter example in Figure 38-1

$$R = X_C = 1 / (2\pi f_C C)$$

so

$$f_C = 1 / (2\pi R C)$$

For the second low-pass filter in example in Figure 38-1

$$R = X_L = 2\pi f_C L$$

so

$$f_C = R / (2\pi L)$$

38.2.1 High-Pass Filter Circuit Example 1

Refer to the circuit in Figure 38-3.

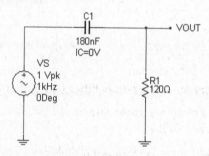

Figure 38-3: High-Pass Filter Circuit Example 1

Calculate the circuit values indicated in Table 38-1 for the frequencies indicated. Recall that for this type of low-pass filter that $f_C = 1 / [2\pi(R1)(C1)]$.

Table 38-1: Calculated Circuit Values for High-Pass Filter Circuit Example 1

Circuit Value	$f = f_C / 10$	$f = f_C$	$f = 10 f_C$
f			
$X_{C1} = 1 / [2\pi f(C1)]$			
$Z_T = \sqrt{R1^2 + X_{C1}^2}$			
$VOUT = VS \times (X_{C1} / Z_T)$			
$VOUT$ (dB) $= -20 \log (VOUT)$			
$\varphi = \tan^{-1}(X_{C1} / R1)$			

Calculate the decades N above f_C and the amplitude A in decibels for each of the frequencies f shown in Table 38-2.

Table 38-2: Calculated Amplitude Values for High-Pass Filter Circuit Example 1

f	$N = \log(f/f_C)$	$A = N \times 20$ dB
20 Hz		
50 Hz		
100 Hz		
200 Hz		
500 Hz		

38.2.2 High-Pass Filter Circuit Example 2

Refer to the circuit in Figure 38-4.

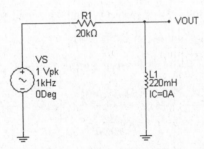

Figure 38-4: High-Pass Filter Circuit Example 2

Calculate the circuit values indicated in Table 38-3 for the frequencies indicated. Recall for this type of low-pass filter that $f_C = R1 / [2\pi(L1)]$.

Table 38-3: Calculated Circuit Values for High-Pass Filter Circuit Example 2

Circuit Value	$f = f_C / 10$	$f = f_C$	$f = 10 f_C$
f			
$X_{L1} = 2\pi(f)(L1)$			
$Z_T = \sqrt{R1^2 + X_{L1}^2}$			
$VOUT = VS \times (X_{L1} / Z_T)$			
$VOUT$ (dB) $= -20 \log(VOUT)$			
$\varphi = \tan^{-1}(R1 / X_{L1})$			

Calculate the decades N below f_C and the amplitude A in decibels for each of the frequencies f shown in Table 38-4.

Table 38-4: Calculated Amplitude Values for High-Pass Filter Circuit Example 2

f	$N = \log(f / f_C)$	$A = N \times 20$ dB
50 Hz		
100 Hz		
200 Hz		
500 Hz		
1 kHz		

38.3 Design Verification

38.3.1 High-Pass Filter Circuit Example 1

1. Open the circuit file Ex38-01.

2. Connect the input of the Bode plotter across **VS** and the output between **VOUT** and ground.

3. Set the Bode plotter settings to those in Table 38-5.

Table 38-5: Bode Plotter Settings

Mode	Horizontal	Vertical
Magnitude	Log F: 1 MHz I: 1 Hz	Log F: 20 dB I: -80 dB
Phase	Log F: 1 MHz I: 1 Hz	Lin F: 95 deg I: −95 deg

4. Measure and record the circuit amplitude and phase readings for each of the frequencies you calculated for Table 38-1 in Table 38-6.

Table 38-6: Measured Circuit Values for High-Pass Filter Circuit Example 1

Circuit Value	$f = f_C / 10$	$f = f_C$	$f = 10 f_C$
f			
Magnitude (dB)			
Phase (degrees)			

Do the measured circuit values in Table 38-6 agree with your calculated values in Table 38-1?

5. Measure the amplitude A for each of the frequencies f shown in Table 38-7 and record the values.

Table 38-7: Measured Amplitude Values for High-Pass Filter Circuit Example 1

f	A
20 Hz	
50 Hz	
100 Hz	

f	A
200 Hz	
500 Hz	

Do the measured circuit values in Table 38-7 agree with your calculated values in Table 38-2?

38.3.2 Low-Pass Filter Circuit Example 2

1. Open the circuit file Ex38-02.

2. Connect the input of the Bode plotter across *VS* and the output between *VOUT* and ground.

3. Set the Bode plotter settings to those in Table 38-8.

Table 38-8: Bode Plotter Settings

Mode	Horizontal	Vertical
Magnitude	Log F: 1 MHz I: 1 Hz	Log F: 20 dB I: −80 dB
Phase	Log F: 1 MHz I: 1 Hz	Lin F: 95 deg I: −95 deg

4. Measure the circuit amplitude and phase readings for each of the frequencies you calculated for Table 38-3 and record them in Table 38-9.

Table 38-9: Measured Circuit Values for High-Pass Filter Circuit Example 2

Circuit Value	$f=f_C/10$	$f=f_C$	$f=10f_C$
f			
Magnitude (dB)			
Phase (degrees)			

Do the measured circuit values in Table 38-9 agree with your calculated values in Table 38-3?

5. Measure the amplitude *A* for each of the frequencies *f* shown in Table 38-10 and record the values.

Table 38-10: Measured Amplitude Values for High-Pass Filter Circuit Example 2

f	A
50 Hz	
100 Hz	
200 Hz	
500 Hz	
1 kHz	

Do the measured circuit values in Table 38-10 agree with your calculated values in Table 38-4?

38.4 Application Exercise

Just as a load can affect the output of a voltage divider, a load can affect the frequency response of a filter circuit. Refer to the loaded filter circuits and their circuit equivalents in Figure 38-5 and Figure 38-6.

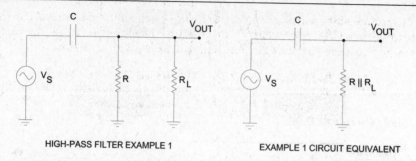

HIGH-PASS FILTER EXAMPLE 1   EXAMPLE 1 CIRCUIT EQUIVALENT

Figure 38-5: Loaded Filter Circuit Example 1 and Circuit Equivalent

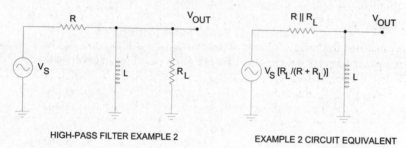

HIGH-PASS FILTER EXAMPLE 2   EXAMPLE 2 CIRCUIT EQUIVALENT

Figure 38-6: Loaded Filter Circuit Example 2 and Circuit Equivalent

As you can see, the circuit equivalents for both loaded circuits have the same form as the original unloaded circuit but with different circuit values. The circuit equivalents for both loaded high-pass filters have a resistance value of $R \parallel R_L$ so that the corner frequency is affected. In addition, the circuit equivalent for the second high-pass filter example, which you can obtain by Thevenizing the circuit, shows that a load resistor will affect the amplitude of the output voltage as well.

For the circuit in Figure 38-7, calculate and record the equivalent resistance and corner frequency in Table 38-11.

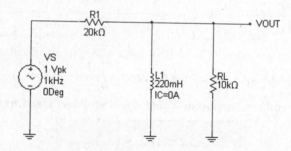

Figure 38-7: High-Pass Filter Circuit with 10 kΩ Load

Table 38-11: Calculated Values for Loaded High-Pass Filter

Circuit Value	Calculated Value
$R_{EQ} = R1 \parallel RL$	
$f_C = R_{EQ} / [(2\pi)(L1)]$	

Open the circuit file Pr38-01. Use the Bode plotter to measure the ideal −3 dB corner frequency for the loaded low-pass filter and record this value in Table 38-12. Note that this is the value that is 3 dB less than the pass-band amplitude, rather than the value that is equal to −3 dB. To find this value, use the cursor right-click menu to find the maximum Y value and then search for the value that is 3 dB less than this value, rather than for –3 dB. For example, if the pass-band amplitude is –10 dB, the –3 dB corner frequency is the frequency for which the amplitude is –10 dB – 3 dB = –13 dB.

Table 38-12: Measured Values for Loaded High-Pass Filter

Circuit Value	Calculated Value
f_C	

How does your measured value of f_C in Table 38-12 compare with your measured value of f_C in Table 38-9?

38.5 Section Summary

This section presented the basic attributes of passive high-pass filters and the Bode plot used to describe the amplitude and phase frequency response of filters. High-pass filters pass frequencies above a specific frequency, called the corner frequency. The high-pass filter blocks frequencies below this corner frequency so that the output rolls off, or decreases, 20 dB per decade. High-pass filters also exhibit a phase shift as the frequencies decrease from passband through the corner frequency and beyond.

39. Band-Pass Filter Circuits

39.1 Introduction

Band-pass filters combine the frequency response of a high-pass and low-pass filter by blocking frequencies below a low corner frequency f_{CL} and above a high corner frequency f_{CH} and passing the frequencies between them. In fact, one way of creating a band-pass filter is to cascade, or connect in sequence, a high-pass filter and a low-pass filter. You can also use a series RLC circuit to create a band-pass filter.

In this section you will:

- Study the amplitude and phase response of passive band-pass filters.

- Investigate the effects of corner frequency separation on the response of passive band-pass filters.

- Use the Multisim software to verify the characteristics of passive band-pass filters.

39.2 Pre-Lab

39.2.1 Cascaded Band-Pass Filter Circuit

Refer to the circuit of Figure 39-1.

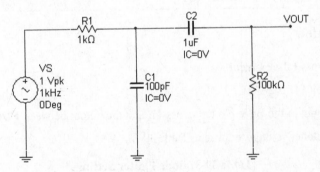

Figure 39-1: Cascaded Band-Pass Filter Example

In this circuit *R1* and *C1* form a low-pass filter that sets the high corner frequency, and *R2* and *C2* form a high-pass filter that sets the low corner frequency. Calculate the low-pass and high-pass corner frequencies for the circuit and record them in Table 39-1.

Table 39-1: Calculated Cascaded Band-Pass Filter Corner Frequencies

Frequency	Calculated Value
$f_{CH} = 1 / [2\pi(R1)(C1)]$	
$f_{CL} = 1 / [2\pi(R2)(C2)]$	

39.2.2 Series RLC Band-Pass Filter Circuit

Refer to the circuit of Figure 39-2.

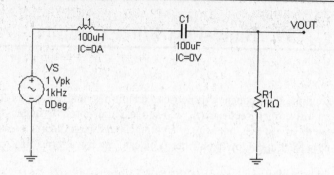

Figure 39-2: Band-Pass Filter Example 2

In this circuit **R1** and **L1** form a low-pass filter that sets the high corner frequency, and **R1** and **C1** form a high-pass filter that sets the low corner frequency. Calculate the low-pass and high-pass corner frequencies for the circuit and record them in Table 39-2.

Table 39-2: Calculated Series RLC Band-Pass Filter Corner Frequencies

Frequency	Calculated Value
$f_{CH} = R1 \,/\, [2\pi(L1)]$	
$f_{CL} = 1 \,/\, [2\pi(R1)(C1)]$	

39.3 Design Verification

39.3.1 Cascaded Band-Pass Filter Circuit

1. Open the circuit file Ex39-01.

2. Connect the input of the Bode plotter across **VS** and the output between **VOUT** and ground.

3. Set the Bode plotter settings to those in Table 39-3.

Table 39-3: Bode Plotter Settings

Mode	Horizontal	Vertical
Magnitude	Log F: 100 MHz I: 1 mHz	Log F: 0 dB I: −60 dB
Phase	Log F: 100 MHz I: 1 mHz	Lin F: 95 deg I: −95 deg

4. Set the Mode setting to "Magnitude".

5. Simulate the circuit. You should see a magnitude Bode plot similar to that in Figure 39-3.

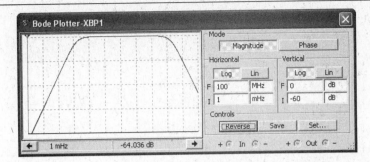

Figure 39-3: Cascaded Band-Pass Filter Magnitude Bode Plot

6. Measure the −3 dB frequencies for the low and high filter corners and record them in Table 39-4.

7. Set the Mode setting to "Phase". You should see a phase Bode plot similar to that in Figure 39-4.

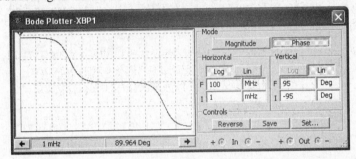

Figure 39-4: Cascaded Band-Pass Filter Phase Bode Plot

Note that the band-pass filter has two phase transitions that correspond to the low and high corner frequencies. The first transition, corresponding to the RC high-pass filter, shifts the phase from 90° to 0°, and the second transition, corresponding to the RC low-pass filter, shifts the phase from 0° to −90°.

8. Measure the phase for the low and high corner frequencies that you measured in Step 6 and record them in Table 39-4.

Table 39-4: Measured Cascaded Band-Pass Filter Corner Values

Filter Value	Measured Frequency	Measured Phase
f_{CH}		
f_{CL}		

Do your measured corner frequencies in Table 39-4 agree with your calculated values in Table 39-1?

39.3.2 Series RLC Band-Pass Filter Circuit

1. Open the circuit file Ex39-02.

2. Connect the input of the Bode plotter across *VS* and the output between *VOUT* and ground.

3. Set the Bode plotter settings to those in Table 39-5.

Table 39-5: Bode Plotter Settings

Mode	Horizontal	Vertical
Magnitude	Log F: 100 MHz I: 1 mHz	Log F: 0 dB I: −60 dB
Phase	Log F: 100 MHz I: 1 mHz	Lin F: 95 deg I: −95 deg

4. Set the Mode setting to "Magnitude".

5. Simulate the circuit. You should see a magnitude Bode plot similar to that in Figure 39-5.

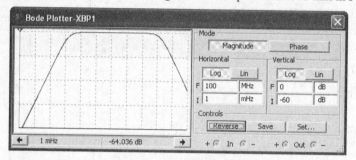

Figure 39-5: Cascaded Band-pass Filter Magnitude Bode Plot

6. Measure the −3 dB frequencies for the low and high filter corners and record them in Table 39-6.

7. Set the Mode setting to "Phase". You should see a phase Bode plot similar to that in Figure 39-6.

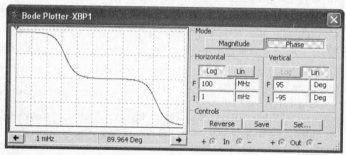

Figure 39-6: Cascaded Band-pass Filter Phase Bode Plot

Note that the bandpass filter has two phase transitions that correspond to the low and high corner frequencies. The first transition, corresponding to the RC high-pass filter, shifts the phase from 0° to 90°, and the second transition, corresponding to the RL low-pass filter, shifts the phase from 0° to −90°.

8. Measure the phase for the low and high corner frequencies that you measured in Step 6 and record them in Table 39-6.

Table 39-6: Measured Cascaded Band-Pass Filter Corner Values

Filter Value	Measured Frequency	Measured Phase
f_{CH}		
f_{CL}		

Do your measured corner frequencies in Table 39-6 agree with your calculated values in Table 39-2?

39.4 Application Exercise

A fairly obvious requirement for the corner frequencies of a band-pass filter is that the low corner frequency must be lower than the high corner frequency. A less obvious requirement is that the corner frequencies must have a minimum separation to avoid unwanted interaction between the two filters that comprise the band-pass filters.

Open the circuit file Pr39-01. For each value of *C2* listed in Table 39-7, calculate f_{CL}. Then measure the corner frequencies for the bandpass filter and record these values in Table 39-7.

Table 39-7: Practice Circuit Values

C2	Calculated f_{CL}	Measured f_{CL}	Measured f_{CH}
1 µF			
100 nF			
10 nF			
1 nF			
100 pF			
10 pF			
1 pF			

For what value of *C2* does your measured value of f_{CL} differ from your calculated value by more than 2%?

For what value of *C2* does your measured value of f_{CH} differ from its initial value by more than 2%?

39.5 Section Summary

This section presented the basic characteristics of passive band-pass filters, which combine the frequency response of passive high-pass and low-pass filters. Band-pass filters pass frequencies within a certain frequency band between a low corner frequency and high corner frequency. Band-pass filters exhibit two phase shifts that correspond to these two corner frequencies, resulting in a total 180° phase shift as frequencies move from the low frequencies through the low corner frequency, pass band, high corner frequency, and beyond. This extreme phase shift can sometimes lead to unwanted oscillation in circuits that use RC and RL configurations.

40. Band-Stop Filter Circuits

40.1 Introduction

Band-stop filter circuits, also called band-reject filter circuits, are the opposite of band-pass filter circuits. Just as band-pass filter circuits pass frequencies within a band defined by low and high corner frequencies, band-stop filter circuits block, or reject, frequencies within a defined band. This creates a trough or "notch" in the magnitude Bode plot, which is why this filter is sometimes called a notch filter.

In this section you will:

- Study the amplitude and phase response of series RLC band-stop filters.

- Investigate the effects of loading on the frequency response of series RLC band-stop filters.

- Use the Multisim software to verify the characteristics of series RLC band-stop filters.

40.2 Pre-Lab

A simple way to create a band-stop filter is to exchange the resistive and reactive elements in a series RLC band-pass filter. Refer to the circuit in Figure 40-1.

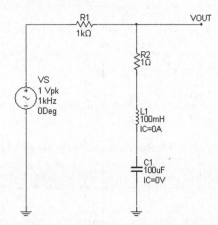

Figure 40-1: Series RLC Band-Stop Filter

Note that this series RLC circuit has two resistors as opposed to a single resistor in the series RLC band-pass filter. This is because the minimum amplitude of the filter in the stop band is not arbitrarily small. *R1* and *R2* are a voltage divider whose ratio sets the value of this minimum stop-band amplitude as follows:

$$VOUT = VS \times [R2 / (R1 + R2)]$$

$$A_{MIN} \text{ (dB)} = 20 \log (VOUT / VS)$$

$$= 20 \log \{[VS \times (R2 / (R1 + R2))] / VS\}$$

$$= 20 \log [R2 / (R1 + R2)]$$

At low frequencies *L1* acts as a short so that *R1* and *C1* determine the low frequency pass-band response and set the low frequency corner at

$$f_{CL} = 1 / [2\pi(R1 + R2)(C1)]$$

At high frequencies *C1* acts as a short so that *R1* + *R2* and *L1* determine the high frequency pass-band response and set the high frequency corner at

$$f_{CH} = (R1 + R2) / [2\pi(L1)]$$

For the circuit in Figure 40-1 calculate f_{CL}, f_{CH}, and the minimum stop-band amplitude in decibels and record your values in Table 40-1.

Table 40-1: Calculated Band-Stop Filter Values

Filter Value	Calculated Value
$f_{CL} = 1 / [2\pi(R1 + R2)(C1)]$	
$f_{CH} = (R1 + R2) / [2\pi(L1)]$	
A_{MIN} (dB) $= 20 \log [R2 / (R1 + R2)]$	

Calculate and record the minimum stop-band amplitude for each set of values of **R1** and **R2** in Table 40-2.

Table 40-2: Calculated Minimum Stop-Band Amplitude

R1	R2	A_{MIN} (dB) $= 20 \log [R2 / (R1 + R2)]$
990	10	
970	30	
950	50	
930	70	
910	90	

40.3 Design Verification

Open the circuit file Ex40-01. Use the Bode plot to measure f_{CL}, f_{CH}, and the minimum stop-band amplitude and record the values in Table 40-3.

Table 40-3: Measured Band-Stop Filter Values

Filter Value	Measured Value
f_{CL}	
f_{CH}	
A_{MIN} (dB)	

Do your measured values in Table 40-3 agree with your calculated values in Table 40-1?

Use the Bode plotter to examine the phase plot for the notch filter.

How many phase transitions does the filter exhibit?

How does the phase change over frequency?

What is the total phase shift for the filter as the frequency moves from the lower pass band through the stop band region and into the upper pass band?

Change the values of **R1** and **R2** to each of the set of values in Table 40-4, measure the minimum stop-band amplitude for each, and record the measured values.

Table 40-4: Calculated Minimum Stop-Band Amplitude

R1	R2	A_{MIN} (dB) = 20 log [R2 / (R1 + R2)]
990	10	
970	30	
950	50	
930	70	
910	90	

Do your measured values in Table 40-4 agree with your calculated values in Table 40-2?

40.4 Application Exercise

One issue with the series RLC band-stop filter circuit is that the frequency response is fairly sensitive to loading. Refer to the loaded band-stop filter and its circuit equivalent in Figure 40-2.

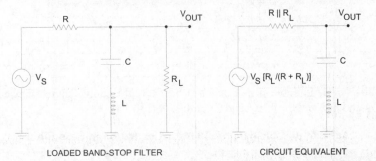

LOADED BAND-STOP FILTER   CIRCUIT EQUIVALENT

Figure 40-2: Loaded Series RLC Band-Stop Filter Circuit

With R_L connected to the circuit, total resistance in the RLC circuit changes from R to $R \parallel R_L$ so that the corner frequencies change to

$$f_{CL} = 1 / [2\pi(R \parallel R_L)(C)]$$

and

$$f_{CH} = (R \parallel R_L) / (2\pi L)$$

Since adding a resistor in parallel always decreases the original resistance this means that f_{CL} will increase and f_{CH} will decrease, moving the corner frequencies closer together. In addition, the maximum pass-band value of V_{OUT} decreases to $V_S [R_L / (R + R_L)]$ so that

$$
\begin{aligned}
A_{PASS} \text{ (dB)} \quad &= 20 \log (V_{OUT} / V_S) \\
&= 20 \log (\{V_S [R_L / (R + R_L)]\} / V_S) \\
&= 20 \log [R_L / (R + R_L)]
\end{aligned}
$$

Because of this, you must measure the -3 dB corner frequencies when the amplitude is 3 dB less than the pass-band amplitude, rather than -3 dB.

Refer to the circuit in Figure 40-3.

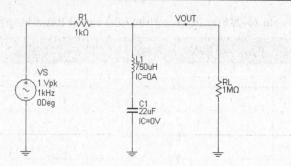

Figure 40-3: Loaded Series RLC Band-Stop Filter Practice Circuit

Calculate the nominal (unloaded) circuit values for the circuit and record them in Table 40-5. Note that **VOUT = VS** in the pass band.

Table 40-5: Unloaded Circuit Values

Circuit Value	Calculated Value
$f_{CL} = 1 / [2\pi(R1)(C1)]$	
$f_{CH} = R1 / (2\pi L1)$	
A_{PASS} (dB)	

For each of values of R_L in Table 40-6, calculate and record the equivalent resistance R_{EQ}, the corner frequencies f_{CL} and f_{CH}, and the pass-band amplitude A_{PASS} (dB).

Table 40-6: Calculated Values for Loaded Band-Stop Filter

RL	$R_{EQ} = R1 \parallel RL$	f_{CL}	f_{CH}	A_{PASS} (dB)
100 kΩ				
50 kΩ				
20 kΩ				
10 kΩ				
5 kΩ				
2 kΩ				
1 kΩ				

Open the circuit file Pr40-01. Change the load resistor **RL** to each of the values in Table 40-7, and measure and record the −3 dB corner frequencies and pass-band amplitude for each. Remember to measure the corner frequencies when the amplitude is −3 dB relative to the pass-band amplitude.

Table 40-7: Measured Values for Loaded Band-Stop Filter

RL	$R_{EQ} = R1 \parallel RL$	f_{CL}	f_{CH}	A_{PASS} (dB)
100 kΩ				
50 kΩ				
20 kΩ				
10 kΩ				
5 kΩ				
2 kΩ				
1 kΩ				

Do your measured values in Table 40-7 agree with your calculated values in Table 40-6?

What is the minimum load resistance for which the filter values are within 10% or 0.85 dB of the unloaded values in Table 40-5?

40.5 Section Summary

This section presented the basic characteristics of the series RLC band-stop filter, which is similar to the series RLC band-pass filter. This filter allows you to set the minimum stop-band amplitude, but is sensitive to loading effects on the frequency corners and pass-band amplitude.